COURSES AND LECTURES - No. 264

NEW PHYSICAL TRENDS IN EXPERIMENTAL MECHANICS

EDITED BY

J.T. PINDERA

UNIVERSITY OF WATERLOO

Springer-Verlag Wien GmbH

ISBN 978-3-211-81630-1   ISBN 978-3-7091-4344-5 (eBook)
DOI 10.1007/978-3-7091-4344-5

5. Check of Validity of Assumptions 106
    5.1. Assumptions of Continuum, Homogeneity and Isotropy 107
    5.2. Assumption about a Known Initial Field 111
    5.3. Experiments for Determination of the Constitutive Equations, Notably the $\sigma_{ij} - e_{ij} - t - \theta$ relationship 112
    5.4. Determination of Boundary Conditions 123
6. References 131

Advanced Theoretical and Experimental Analysis of Plates and Plates in Contact
by K.H. Laermann

1. Introduction 135
2. Mathematical Model of an Extended Plate Theory 139
3. Photoelasticity 154
4. Moirè Methods 177
5. Combined Method 186
6. Strain Gage Method 190
7. Final Remarks 196
8. References 196

Foundations of Experimental Mechanics: Principles of Modelling Observation and Experimentation
by J.T. Pindera

Preface 199
Part I
1. Introduction: Survey, Approach 203
2. Contemporary Developments in Experimental Mechanics 211
3. Modelling of Reality 225
4. Modelling in Mechanics 243
5. General Patterns of Determination of Response of Real Bodies 254
6. Elements of Information Producing Process 268
7. Summary of Part I 288
Part II
8. Coupled Rheological Responses of Materials Used in Model Mechanics 299
9. Stress State in a Circular Disk — Contradictions Between the Analytical and Experimental Results 312
10. Non-Rectilinear Light Propagation in a Stressed Body 314
11. Isodyne Photoelasticity 317
12. Limits of Conventional Ultrasonic Techniques 319
References 321

Flow Visualization

by C. Truchasson

Preface 329
1. Generalities 331
2. Material Tracer Techniques 334
3. Energy Addition Used as a Tracer Technique 342
4. Optical Methods 346
5. Analogical Systems in Flow Visualization 353
6. Recording and Illumination Technique in Flow Visualization 358
Conclusion 362
References 364

EXPERIMENTAL MECHANICS APPLIED TO THE ACCELERATED

CHARACTERIZATION OF POLYMER BASED COMPOSITE MATERIALS

H. F. Brinson, Professor
Department of Engineering Science and Mechanics
Virginia Polytechnic Institute and State University
Blacksburg, Virginia 24061

Introduction

Composite materials have been used as structural materials almost
since the beginning of recorded civilization. Examples from the mud
bricks with straw used by the early Inca and Egyptian societies to the
concrete and plywood used by modern societies are well known to all.
However, the purpose here is to discuss a new breed of resin matrix
composite materials. The two types which will be considered herein are
continuous fiber graphite/epoxy (G/E) resin laminated composites, often
referred to as advanced composite materials, and a chopped fiberglass/
polyester resin composite, often referred to as sheet molding compound
(SMC).

The principal engineering advantages of these materials arise due
to their high strength to weight ratios. Their use in the design of

automobiles, trucks, planes, rockets, etc., frequently results in light structures with high efficiency; i.e., increased payloads with less fuel consumption.

The anisotropic and heterogeneous nature of composites presents unique opportunities for the use of the methods of experimental mechanics to measure moduli and failure and fracture parameters and their mechanisms. Added to these already formidable problems is sensitivity to the environmental parameters of time, temperature and moisture. Another factor which creates further complications for the experimentalist is statistical variability.

It should be noted that, especially for composite materials, analysis and experiment cannot be separated. Analysis depends upon the material properties which can only be found in the laboratory. On the other hand, each laboratory test represents an attempt to duplicate a well posed boundary value problem for which a good analytical solution is available. Because of heterogeneity and anisotrophy, we must examine anew all of our well-established test techniques which have evolved to determine properties of homogeneous and isotropic materials. This is no trivial task and the need for combined analytical and experimental expertise cannot be over-emphasized.

The objective herein is to present information pertinent to the time dependent viscoelastic behavior of a particular G/E material - T300/934 and an SMC chopped fiber composite. The intent is to examine ways and means to evaluate experimentally and analytically the time dependent deterioration of matrix dominated moduli and strengths. Also,

the elements of an accelerated characterization plan will be presented
and documented.

The discussion of viscoelastic effects and their accelerated
characterization will, of necessity, touch on a number of subjects related
to composite materials and the application of experimental mechanics
thereto. For example, the properties of a single ply are needed for the
use of lamination theory. Thus, a brief description of lamination theory
and ways and means to determine the properties of a lamina will be discussed. Failure and fracture will also be touched on briefly as an
introduction to understanding the more complex problems of delayed viscoelastic failure and fracture of composite materials. However, the main
thrust of what is to be presented here is related to viscoelastic testing
and analysis procedures as applied to both continuous and chopped fiber
polymer matrix composites. As a short article of this type cannot
possibly contain a complete overview of all that has been accomplished
relative to composite materials over the last two decades the reader is
referred to several excellent texts on the subject.[1-8]

Accelerated Characterization Plan

Today's energy-oriented society tends to require light but strong
structural materials operating at high stress levels for a prolonged
time under possible hostile environmental conditions. Advanced polymeric
composites seem to be a perfect material because they are light, strong
and can be tailored for a particular application. That is, they can be
designed with the strong and stiff fibers in the directions of highest
stress while the relatively weak and compliant polymeric matrix operates

under low stress levels. However, the matrix is an important structural
component as it serves to transfer the applied stress from fiber to fiber
and from ply to ply. For this reason, the matrix, and hence matrix dominated properties, plays an important role in structural design and/or
behavior of a composite material and cannot be ignored.

Matrix dominated moduli and strength properties of polymer based
composite laminates are time dependent or viscoelastic and are sensitive
to environmental conditions such as temperature and humidity. Because of
this fact, the long term integrity of a composite structural component is
an important consideration in the initial design process. Therefore, how
viscoelastic matrix dominated modulus (compliance) and strength properties
vary with time over the design life time is necessary input to the
initial design process. As many structural components are designed for
years of service, property variations over years is often needed. Obviously, long term testing equivalent to the lifetime of a structure is
impractical and undesirable. The alternative is to develop analytical
and experimental methods which can be successfully used for extrapolation.

The accelerated characterization procedure which was developed for
polymer based composite laminates several years ago by Brinson[9] is based
upon the well-known time-temperature superposition principle (TTSP) for
polymers and the widely used lamination theory for composite materials.
A block diagram illustrating the basic details of the plan is shown in
Fig. 1. The generic idea was to develop a method by which the time
dependent deterioration of laminate moduli (compliances) and strength
could be calculated from the results of a minimum number of tests.

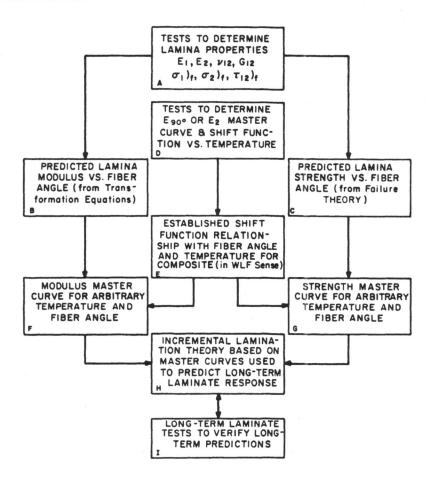

Fig. 1. Flow chart for proposed laminate accelerated characterization
         and failure predictions procedures.

Hopefully, the amount of testing would only be minimally greater than

that required for normal quality control and/or basic property determi-

nation procedures.[10]

Advanced composite laminates are most frequently designed using

laminatation theory. This theory allows the calculation of the proper-

ties of a general laminate from the knowledge of the behavior of a

single lamina or ply. The stress-strain properties of a single ply may
be found from constant strain-rate tests on unidirectional laminates and
are normally routinely obtained when a general laminate is made. Thus,
the accelerated characterization plan shown in Fig. 1 assumes that lamina
stress-strain properties from zero load to failure are known as indicated
by item A. Methods to determine these ramp loaded properties will be
discussed in a subsequent section.

The transformation equation for the moduli of orthotropic materials
has been shown to be valid for unidirectional laminates.[10] Also, various
orthotropic failure theories have been shown to be valid for unidirectional laminates.[11] Therefore, from item A modulus and strength
properties as a function of fiber angle are known as indicated in items
B and C.

Before time dependent properties of a general laminate can be predicted, knowledge of the time dependent behavior of a single ply is necessary. For this reason, the constant strain-rate behavior known from
routine tests as given in items A, B and C are insufficient for viscoelastic predictions. To perform long term creep or relaxation tests to
determine the necessary lifetime-information is impractical and does not
satisfy the objective of making long term predictions from a minimum
number of tests conducted in a short time.

The fundamental concept employed to overcome the above obstacle was
to use the well-known TTSP principle to produce a modulus (compliance)
master curve for a single fiber orientation as typified by item D of Fig.
1. The TTSP principle applied to composite laminates requires the

conduction of short term creep tests on a unidirectional laminate at
various temperatures. These can likely be performed in a single day or a
few days at most. Next, the plan of Fig. 1 assumed that either an Arhenius
or WLF type equation could be modified to predict the variation of shift
function with fiber angle for a single lamina without further testing as
represented by item E. If such could be done, then the results of D and
E combined with the information of A and B would produce the modulus
(compliance) master curves of item F by simple scaling procedures without
additional testing.

Delayed failure predictions, of course, require knowledge of time
dependent strength properties. The determination of such properties often
require large amounts of testing over a prolonged period of time. Manu-
facturers would surely be reluctant to include an extensive testing
program for routine quality control and property determination procedures.
To avoid an extensive creep rupture testing program, the assumption was
made that strength master curves were of the same shape as modulus (compliance) master curves for any particular fiber angle. From this assumption lamina strength master curves as a function of fiber angle and
temperature of item G could be determined from items C and F again by
simple scaling procedures.

Given the master curves of F and G, an incremental lamination theory
was to be developed to predict the long term modulus and strength
properties of a general laminate. The results were to be compared with
experiments as specified by item I.

Commentary on Accelerated Characterization Plan

The plan detailed in the previous section, if proven valid, would give a manufacturer of polymer based composite structural components a method by which time dependent failures could be anticipated and avoided. Failures might be either the degradation of moduli with time, creep ruptures, matrix delaminations, ply separation, etc. The important feature is that only a single master curve and associated shift function relationship for a single lamina need be produced in addition to ordinarily or routinely performed stress-strain properties. All necessary information could be produced in a very short time, perhaps a few days, and predictions on the order of years could be made.

The accelerated characterization plan, as previously discussed, requires knowledge of methods to determine the properties of a single ply or lamina, failure laws and, of course, information on the fracture behavior of laminates. These concepts, even without the inclusion of viscoelastic effects, are not that well developed for composite structures. As a result the next several sections represent brief introductions to lamina and laminate stress-strain or constitutive characterization procedures as well as the application of failure and fracture theories to composite laminates.

Perhaps at this point it would be appropriate to state explicitly what is meant by these three necessary items. Herein, constitutive properties is taken to mean the measured relationships between stress and strain and their analytical characterization from initial loading (zero stress and strain) until separation (fracture). Constitutive properties include moduli and Poisson's ratios which may vary with strain level and with strain rate. That is, linear elasticity is not necessarily assumed.

Failure properties refers to the measurement of critical strengths and their analytical quantification via a failure, yield or strength theory. Such properties refer to materials containing only natural flaws where the size of the flaws are insignificant when compared to other material or constituent dimensions. Fracture, on the other hand, refers to the identification of a measured critical strength associated with a measured critical size flaw, notch or crack together with an analytical relation between strength and flaw size.

With these definitions it is intended to indicate that when dealing with laminated composites, experiment and analysis must and should go hand-in-hand. The fundamental idea is that if generic properties of the constituents of a laminate (i.e., the matrix and the fibers or a single ply) are available, then the properties of the laminate can be predicted by analytical techniques. This is in contrast to the usual design procedures where prediction of the behavior of a metallic alloy is rarely attempted knowing only the behavior of the ingredients.

Lamina and Laminate Stress-Strain Characterization

Polymer based fiber reinforced composites are, in general, aniso-tropic and inhomogeneous materials. Except possibly for the rule of mixtures approach to stiffness, inhomogeneity is seldom accounted for. Most analytical techniques for the stress-strain characterization of laminated composites is based upon linear anisotropic elasticity. Such a constitutive equation can be written as,

$$\underset{\sim}{\varepsilon} = \underset{\sim}{S} \, \underset{\sim}{\sigma} \tag{1}$$

where $\underset{\sim}{\varepsilon}$, $\underset{\sim}{S}$, and $\underset{\sim}{\sigma}$ are the strain, compliance and stress tensors, respectively. Alternately equation (1) could be written as,

$$\underset{\sim}{\sigma} = \underset{\sim}{Q} \, \underset{\sim}{\varepsilon} \tag{2}$$

where $\underset{\sim}{Q}$ is the stiffness tensor. The quantities $\underset{\sim}{S}$ and $\underset{\sim}{Q}$ must be experimentally determined and represent a minimum of 21 independent elastic constants for general anisotropy. Obviously, if general laminates were considered to be anisotropic, extensive test programs would have to be performed to obtain only these necessary constants. Further, obvious difficulties would exist if heterogeneity were included.

Often, linear orthotropic elastic constitutive equations are used to describe the behavior of individual plies or laminae or, sometimes, the entire laminate. The former is the foundation of lamination theory.[1-7] The latter can be used for unidirectional laminates and as a first approximation to balanced symmetric general laminates.

For orthotropic materials the number of necessary constants is reduced from 21 to nine.[8] For the case of plane stress the number of constants is further reduced to four. Often a lamina is considered to be orthotropic and in a state of plane stress. Thus, equation (1) for a lamina can be expressed as,[8]

$$\varepsilon_x = \frac{1}{E_x} \sigma_x - \frac{\nu_{yx}}{E_y} \sigma_y$$

$$\varepsilon_y = \frac{\nu_{xy}}{E_x} \sigma_x + \frac{1}{E_y} \sigma_y$$

$$\varepsilon_z = \frac{\nu_{xz}}{E_x} \sigma_x - \frac{\nu_{yz}}{E_y} \sigma_y \tag{3}$$

$$\gamma_{xy} = \frac{1}{G_{xy}} \tau_{xy}$$

where $\sigma_z = \tau_{xz} = \tau_{yz} = 0$, the principal material directions are x and y, the longitudinal and transverse moduli are E_x and E_y respectively, the longitudinal and transverse Poisson's ratios are ν_{xy} and ν_{yx} respectively, and the shear modulus is G_{xy}. Furthermore, $E_x \nu_{yx} = E_y \nu_{xy}$ and the number of independent constants to be experimentally determined is only four.

Because of the possibility of unbalanced and/or unsymmetric layups, most analyses use the properties of a single ply, lamina or orthotropic layer together with orthotropic transformation equations[6,7] to "build up" or calculate the necessary elastic compliances or stiffnesses for either a symmetric or unsymmetric multi-layered laminate. As this laminated plate (lamination) theory can be found in a variety of references[1-7] only an outline of the procedures will be given here. For the sake of brevity, the following is only for the case of symmetric lay-ups. Bending occurs even for the case of simple tension in unsymmetric lay-ups but is beyond the scope intended here.[6]

Consider a unidirectional orthotropic lamina under a state of plane stress in which the principal material directions coincide with the co-ordinate directions x and y, i.e., the x and y are parallel and perpendicular to the fiber direction. Such lamina are called specially orthotropic and for this case the compliances and stiffnesses of equations (1) and (2) can be written as,

$$\underset{\sim}{S} = \begin{bmatrix} S_{11} & S_{12} & 0 \\ S_{21} & S_{22} & 0 \\ 0 & 0 & S_{66} \end{bmatrix} = \underset{\sim}{Q}^{-1} \qquad (4)$$

or

$$\underset{\sim}{Q} = \begin{bmatrix} Q_{11} & Q_{12} & 0 \\ Q_{21} & Q_{22} & 0 \\ 0 & 0 & Q_{66} \end{bmatrix} = \underset{\sim}{S}^{-1} \qquad (5)$$

Substitution of (4) into (1) yields equation (3) after some simplifica-
tion. Relations between $\underset{\sim}{S}$, $\underset{\sim}{Q}$, and the quantities E_x, E_y, ν_{xy} in equations (3) are given in reference 6. Note also that only four of the
quantities $\underset{\sim}{S}$ and $\underset{\sim}{Q}$ in equations (4) and (5) are independent.

If the principal material directions do not coincide with the coordinate axes x and y, equations (4) and (5) are no longer valid, i.e.,
the axes x and y are not parallel and perpendicular to the fiber directions as shown in Fig. 2. This is always the case for an arbitrary
lamina or ply of an angle laminate. For such a lamina equations (1) and
(2) are valid where $\underset{\sim}{S}$ and $\underset{\sim}{Q}$ are given by the matrices,

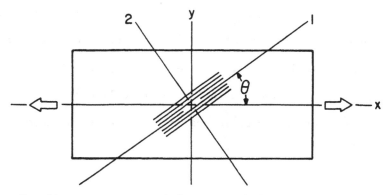

Fig. 2. Coordinate system used for general lamina or ply and for off-
        axis tests.

$$\underset{\sim}{\bar{S}} = \begin{bmatrix} \bar{S}_{11} & \bar{S}_{12} & \bar{S}_{13} \\ \bar{S}_{21} & \bar{S}_{22} & \bar{S}_{23} \\ \bar{S}_{31} & \bar{S}_{32} & \bar{S}_{33} \end{bmatrix} \qquad (6)$$

or

$$\underset{\sim}{\bar{Q}} = \begin{bmatrix} \bar{Q}_{11} & \bar{Q}_{12} & \bar{Q}_{13} \\ \bar{Q}_{21} & \bar{Q}_{22} & \bar{Q}_{23} \\ \bar{Q}_{31} & \bar{Q}_{32} & \bar{Q}_{33} \end{bmatrix} \qquad (7)$$

where the quantities $\underset{\sim}{\bar{S}}$ and $\underset{\sim}{\bar{Q}}$ are transformed compliances and stiffnesses. For example,

$$\bar{S}_{xx} = S_{11} \cos^4\theta + (2S_{12} + S_{66}) \sin^2\theta \cos^2\theta + S_{22} \sin^4\theta \qquad (8)$$

where subscripts refer to the coordinate system shown in Fig. 2. A complete set of the appropriate transformation equations are given in references 6-8.

The laminate stiffnesses are calculated from the transformed laminae stiffnesses, eq. (7), by

$$A = \sum_{k=1}^{n} (\bar{Q})_k (Z_k - Z_{k-1}) \qquad (9)$$

where Z_k represents the distance from the middle surface of the kth layer and $(\bar{Q})_k$ represents the transformed lamina stiffnesses of the kth layer. The laminate compliances are given by,

$$\underset{\sim}{a} = \underset{\sim}{A}^{-1} h \qquad (10)$$

where h is the total thickness of the laminate. In other words, the laminate constitutive equation is,

$$\underset{\sim}{\varepsilon} = \underset{\sim}{a} \underset{\sim}{\sigma} \qquad (11)$$

Thus, the constants, a, necessary to define the behavior of a general laminate can be obtained knowing only the behavior of a single ply. Therefore, using lamination theory, only the properties of a single uni-directional ply need to be determined experimentally. As single plies are quite thin, $\sim$ 0.005 in (0.127 mm), properties are usually determined by testing a multi-layered unidirectional laminate. These are taken as the assumed properties of a single ply and the properties of general laminates are then calculated for each configuration being considered. Obviously, if the properties of the unidirectional laminate do not represent the properties of a single ply or if different defects are present in the manufacture of the general laminate as opposed to the unidirectional laminate, the "calculated" properties are likely to be in error.

Much experimental work has been reported on the stress-strain be-havior of composites.[10,12-23]

Test Methods for Lamina Properties

The orthotropic properties E_x, E_y, ν_{xy} and G_{xy} defined by equation (3) are needed to characterize the behavior of a lamina or ply. The first three of these properties can be determined quite easily by two uniaxial tensile tests. Fig. 2 represents a tensile specimen with fibers at an arbitrary angle with the load direction.

From a tensile test with the load in the fiber direction ($\theta = 0°$), the modulus $E_{11} = E_x$ and Poisson's ratio, $\nu_{12} = \nu_{xy}$, can be determined. From a similar test with the load normal to the fibers ($\theta = 90°$), the modulus $E_{22} = E_y$ and $\nu_{21} = \nu_{yx}$ can be determined.

Determination of the in-plane shear modulus G_{xy} or G_{12} is much more difficult. A variety of shear determination techniques have been proposed. The short beam shear test has been used but shear stress variations through the thickness and different properties in tension and compression has limited its utility.[14] The torsion testing of a thin tube[22] and the picture frame test[23] represent a better approach, but material and equipment expense as well as other difficulties make them generally unattractive. The standard rail shear test[21] also often yields reasonable results. Again, however, material cost as well as the unsymmetric nature of the load (relative to the laminate) represents undesirable features. The symmetric rail shear test[20] avoids the latter difficulty but does not alleviate material requirements and costs.

Use of judiciously chosen tensile tests seems to represent a rational alternative to the above shear testing techniques. The off-axis tensile testing of unidirectional laminates can be used to obtain shear behavior.[13] Also, the tensile testing of $[\pm 45°]_s$ laminates can be used for shear predictions and has been shown to give good results.[18,19]

Off-Axis Tests. One method to obtain the shear modulus of a ply or lamina is to use the results of three tension tests. Using the results of the previously described tensile tests to determine E_{11}, E_{22}, ν_{12} together with the results of a third tension test with the load at an angle to the fiber direction (off-axis tensile test), the shear modulus, G_{12}, can be calculated from the following orthotropic transformation equation,[8]

$$\frac{1}{E_x} = \frac{1}{E_{11}} \cos^4\theta + \left[\frac{1}{G_{12}} - \frac{2\nu_{12}}{E_{11}} \right] \sin^2\theta \cos^2\theta + \frac{1}{E_{22}} \sin^4\theta \qquad (12)$$

in which θ is the fiber direction and E_x is the modulus in the load
direction.

While the above procedure is quite adequate for modulus, it does not
provide a means for the determination of the total stress-strain response.
Another simple procedure using measurements from an off-axis tensile
specimen will yield the entire shear stress-strain curve for a lamina.
All that is required is to measure the strains in three directions on
such a specimen, e.g., in the longitudinal, transverse and 45° directions
which is easily accomplished with an electrical strain gage rosette. By
transforming both the measured strains and the applied stress (ϵ_x, ϵ_y,
$\epsilon_{45°}$ and σ_x), the strains and stresses in the local coordinates can be
found (ϵ_1, ϵ_2, γ_{12}, σ_1, σ_2 and τ_{12}). A plot of τ_{12} vs. γ_{12} yields the
required shear behavior. Chamis[13] has suggested use of a 10° off-axis
specimen for this purpose and has obtained good results with the technique.

Often only rectangular rosettes are used in tensile testing to obtain longitudinal and transverse strains only. Thus, for such cases the
procedure just outlined cannot be used. However, by assuming that equation (12) is valid for tangent values of moduli, it is possible to calculate complete shear stress-strain response using standard lamination
theory in an incremental fashion.[10] It should be noted that the fundamental reason for this procedure for the case where only a rectangular
rosette is used is to be able to obtain a value of γ_{xy}. In other words,
for the off-axis tension test, the principal axis of stress and strain
do not coincide and only an axial stress, σ_x, produces not only an axial

strain, ε_x, but a shear strain, γ_{xy}, as well.

[±45°]$_s$ Tests.   Petit suggested the use of a uniaxial tensile test on a [±45°]$_s$ laminate for the purposes of obtaining the shear stress-strain response of a lamina.[18] He showed that such properties could be obtained from measurements of the tensile load and axial and transverse strains coupled with an incremental lamination theory analysis of the [±45°]$_s$ laminate.   His results were expressed as

$$G_{12} = \frac{2U_1\ E_x}{8U_1 - E_x} \tag{13}$$

where G_{12} was the sought for lamina shear modulus, E_x was the measured tensile modulus of the [45°]$_s$ laminate and U_1 was given by

$$U_1 = \frac{[E_{11} + E_{22} + 2\nu_{21}\ E_{11}]}{8(1 - \nu_{12}\ \nu_{21})} \tag{14}$$

$$\nu_{21} = \frac{\nu_{12}\ E_{22}}{E_{11}} \tag{15}$$

In the latter expressions, the quantities E_{11}, E_{22}, ν_{12} and ν_{21} are the measured principal properties of a unidirectional laminate.   He then expressed the in-plane shearing strain of a lamina as

$$\gamma_{12} = (1 + \nu_{xy})\ \varepsilon_x \tag{16}$$

in which

$$\nu_{xy} = -\frac{\varepsilon_y}{\varepsilon_x} \tag{17}$$

and where ε_x, ε_y and ν_{xy} were the measured tensile properties of a [±45°]$_s$ laminate.   An incremental procedure was used in order to account for the non-linear shear stress-strain response.   The tangent modulus at different strain levels of the [±45°]$_s$ tensile response curve was found and used with equations (13) and (16) to obtain incremental shear

stresses, $\Delta\tau_{12}$, at the various strain increments, $\Delta\gamma_{12}$, from the follow-

ing expression

$$\Delta\tau_{12} = G_{12} \Delta\gamma_{12} \tag{18}$$

where G_{12} was taken from the previous strain level. Thus, the complete

shear stress-strain response was predicted.

Later, Rosen[19] simplified Petit's analysis by noting that the

shearing stress is half of the applied stress for a tensile test in

general and for a $[\pm 45°]_s$ laminate in particular and by further defining

the shearing strain to be the same as equation (16). Rosen's results

were expressed as

$$G_{12} = \frac{\sigma_x}{2(\varepsilon_x - \varepsilon_y)} \tag{19}$$

Good agreement was found between Rosen's and Petit's results.

Rail Shear Tests. A symmetric rail shear test fixture proposed by Sims[20]

for use with $[0°/90°]_s$ laminates is shown in Figure 3. Since the

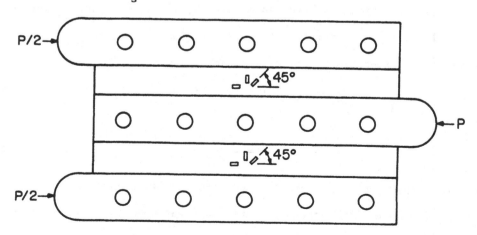

Fig. 3. Symmetric rail shear test fixture showing strain gage
         orientations.

specimen used was symmetric with respect to the applied load, P, the shearing stress, τ_{12}, was expressed as

$$\tau_{12} = \frac{P}{2A} \qquad (20)$$

and the shearing strain as

$$\gamma_{12} = 2\varepsilon_{45°} \qquad (21)$$

where A was the cross-sectional area parallel to the load and the strain was measured at an angle of 45° to the applied load. While Sims did not present a more detailed analysis, laminated plate theory can be used to show that this method does correspond to intralamina shear stress-strain response. As a $[0°/90°]_s$ laminate can be considered to be orthotropic, its constitutive relation can be expressed as

$$\begin{Bmatrix} \varepsilon_x \\ \varepsilon_y \\ \gamma_{xy} \end{Bmatrix} = \begin{bmatrix} S_{11} & S_{12} & 0 \\ S_{12} & S_{22} & 0 \\ 0 & 0 & S_{66} \end{bmatrix} \begin{Bmatrix} N_x \\ N_y \\ N_{xy} \end{Bmatrix} \qquad (22)$$

where N_x, N_y and N_{xy} are the applied loads and ε_x, ε_y and γ_{xy} are the laminate strains with respect to the global axes. The $[S_{ij}]$ matrix is the compliance for the $[0°/90°]_s$ laminate. When only a shear load, N_{xy}, is applied, equation (22) reduces to

$$\gamma_{xy} = S_{66} N_{xy} = \frac{\tau_{xy}}{G_{12}} \qquad (23)$$

as N_{xy} has a dimension of force per unit length and

$$S_{66} = \frac{1}{G_{12} t} \qquad (24)$$

for a $[0°/90°]_s$ laminate. For this laminate, the principal stress and strain axes do coincide. Thus, equation (23) can be expressed as

$$G_{12} = \frac{\tau_{12}}{\gamma_{12}} \qquad (25)$$

or it can be expressed in terms of the experimental quantities of Sims by substituting equations (20) and (21) into (25). The resulting equation can be written as

$$G_{12} = \frac{P}{4A\varepsilon_{45°}} \qquad (26)$$

Test Procedures. Specimen machining needs to be performed carefully to avoid damage to the brittle fibers. A safe method is to use a diamond impregnated saw with a high speed, a low feed and a mister for coolant and lubrication purposes. Specimens should be dried and dessicated to control and maintain a constant moisture content. Variations in moisture content can seriously affect test results.[24]

End tabs are needed for most cases to assure failure occurs within the test section. Because of the random nature of composite test results, a sufficient number of replicates should be used.

All the above questions regarding machining, conditioning, tabbing, replicates, etc., are under close scrutiny by ASTM's D-30 committee. Standard procedures may be found in ASTM Standards, Volume 36. However, procedures are constantly changing and the reader would be best advised to check current literature before beginning an elaborate test program. Measurement of strains can be accomplished using extensometers or electrical strain gages. For basic property determination the writer prefers electrical strain gages as rosettes can be used to determine all the necessary strains.

The measurement of viscoelastic effects often need to be made at elevated temperatures which poses a more stringent requirement than some other cases. Because of difficulties encountered in having strain gages malfunction, the writer and his students have generally used the following strain gage installation procedures for a particular T300/934 graphite epoxy laminate.[25,26]

1. Light sanding of the specimen surface using 220, 320, and 400 grit paper.

2. Cleaning of the sanded surface with propanol until a clean tissue can be wiped over the area without picking up graphite.

3. Baking of the specimen for one hour at 250°F to drive out any absorbed propanol.

4. Installation of the gage using recommended Micromeasurements procedure for M-Bond 610.

5. Curing of the gage installation for two hours at 250°F (this reduces bubble formation) followed by one hour at 350°F.

6. Installation of lead wires, tabs, etc.

7. Post cure for two hours at 430°F.

8. Application of GE RTV 3140 over the gage area for protection.

In addition to the above preparation procedures, care should be taken to account for strain gage heating effects. For this reason 350 Ω gages and low excitation levels (1 to 2 V) are recommended.

In certain situations, when testing composites, transverse sensitivity effects on electrical strain gages must be considered.

<u>Test Results</u>. Stress-strain curves for a particular T300/934 G/E laminate

for the angles 0°, 15°, and 90° as well as the axial strain vs. trans-

verse strain for the 0° laminate are shown in Figures 4 and 5. All the

curves shown are the computer conditioned results of three replicates as

previously described.[10]   The bilinear stiffening behavior of the $[0°]_{8s}$

laminate is apparent.

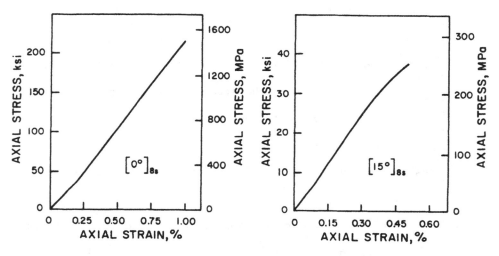

Fig. 4.   Tensile behavior of unidirectional T300/934 G/E laminates.

Initial tensile moduli and ultimate strengths of uniaxial laminates

with the load at various angles to the fiber direction are shown in Fig.

6. Equation (12) is shown superimposed on the moduli results in Fig. 6

with the various properties evaluated from the 0°, 15°, and 90° tests.

This orthotropic transformation equation fits the data extremely well

when the G_{12} value obtained from the 15° data is used and tends to vali-

date the use of an orthotropy assumption for unidirectional materials.

Though not shown, a Tsai-Hill or other failure criteria could be used

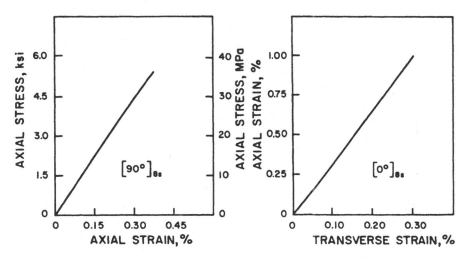

Fig. 5. Tensile behavior of unidirectional T300/934 G/E laminates.

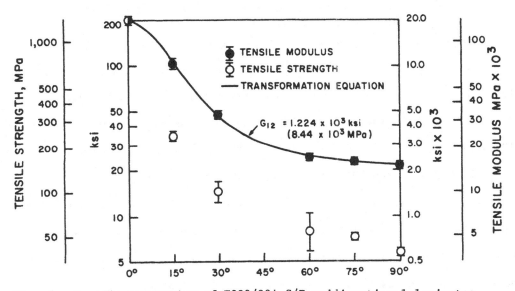

Fig. 6. Tensile properties of T300/934 G/E unidirectional laminates.

to accurately characterize ultimate strength data.[6,11,26]

Shear stress-strain curves are shown in Fig. 7 which were produced

by the procedures discussed in an earlier section. Considerable

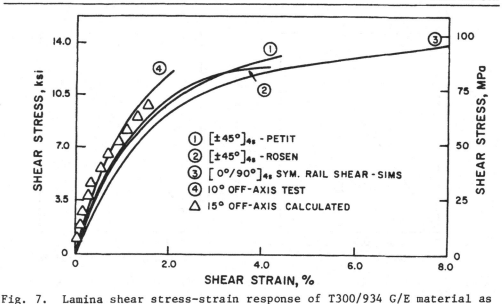

Fig. 7.   Lamina shear stress-strain response of T300/934 G/E material as
          determined by different methods.

differences between the various shear results are apparent.   The two off-

axis shear curves agree with each other quite well but defer drastically

from the other methods.   The fracture stress for the off-axis test

depends heavily on the fiber angle as indicated by Fig. 6.   Obviously a

10° angle is better in this regard than the 15° angle.   Otherwise, the

general shapes of the 10° and the 15° curves are quite similar.

The symmetric rail shear tests results indicate a much higher value

of fracture strain but a lower value of initial modulus.   The $[\pm 45°]_s$

results are intermediate to the off-axis and rail shear tests.   In other

words, shear results from $[\pm 45°]_{4s}$ or $[0°/90°]_{4s}$ laminates appear to be

much more ductile in nature than those from off-axis tests.   This is

reasonable inasmuch as general laminates obviously contain interlamina

as well as intralamina shear response.   In fact, Yeow et al.[10] stated

that post-examination of the off-axis specimens revealed a predominately brittle failure between fibers while similar examination of the $[\pm 45°]_{4s}$

tensile and $[0°/90°]_{4s}$ rail shear specimens revealed extensive delamina-

tion and interply failures prior to separation.

Time Independent Failure Theories

   Numerous failure criteria are presently available[11] which can be

classified as either having independent or dependent failure modes. The

maximum stress or maximum strain criterion are ones with independent

failure modes. The criteria proposed by Ashkenazi,[27] Hill[28] and Puppo

and Evensen[29] are examples of failure theories with dependent failure

modes. Between the two classes, it would seem that those with dependent

failure modes would be more appropriate for composite materials. Experi-

mental results tend to confirm this assumption.[22]

   Ashkenazi's theory assumes a macroscopic continuum in which the

strength properties are fourth order tensors and in which environmental

effects are neglected. His criterion for a plane orthotropic material

(or lamina) under uniaxial load in an arbitrary direction is

$$\frac{1}{\sigma_x} = \frac{\cos^4\theta}{X} + \left(\frac{4}{X_{45}} - \frac{1}{X} - \frac{1}{Y}\right)\sin^2\theta\,\cos^2\theta\,\frac{\sin^4\theta}{Y} \tag{27}$$

where σ_x is the applied normal stress making an angle θ with the prin-

cipal material direction (1 or fiber direction) and X, Y and X_{45} are the

tensile strengths along, transverse to and at 45° to the principal

material direction respectively. While equation (27) is for uniaxial

loading only, other stress states are considered by the author.

   Tsai[30] modified Hill's failure criterion for an orthotropic lamina

under a generalized state of plane stress to be,

$$\left(\frac{\sigma_1}{X}\right)^2 + \left(\frac{\sigma_2}{Y}\right)^2 - \frac{\sigma_1\sigma_2}{X^2} + \left(\frac{\tau_{12}}{S}\right)^2 = 1 \tag{28}$$

where the stresses $\sigma_1\sigma_2$, τ_{12} are along the material axes, S is the in-plane shear strength and X and Y are as previously defined.

A more general theory due to Puppo and Evensen can be expressed as

$$\left(\frac{\sigma_1}{X}\right)^2 - \gamma\left(\frac{X}{Y}\right)\left(\frac{\sigma_1}{X}\right)\left(\frac{\sigma_2}{Y}\right) + \left(\frac{\sigma_2}{Y}\right)^2 + \left(\frac{\tau_{12}}{S}\right)^2 = 1 \tag{29}$$

or

$$\gamma\left(\frac{\sigma_1}{X}\right)^2 - \gamma\left(\frac{Y}{X}\right)\left(\frac{\sigma_1}{X}\right)\left(\frac{\sigma_2}{Y}\right) + \left(\frac{\sigma_2}{Y}\right)^2 + \left(\frac{\tau_{12}}{S}\right)^2 = 1 \tag{30}$$

where $\gamma = \frac{3S^2}{XY}$ is an interaction factor and the other quantities are as previously defined. Further, the authors indicated that their theory could be adapted to a wide range of materials by writing the interaction factor as $\gamma = \left(\frac{3S^2}{XY}\right)^n$. The above equations are for orthotropic materials but the general theory is not so limited. Fig. 8 shows the comparison between theory and experiment temperatures for a G/E material.[26]

The failure theories given above are used in connection with lamination theory to determine when each ply fails and when the laminate fails.[6,7] The difficult question encountered after first ply failure (FPF)[31] is how unloading occurs. Petite and Waddoups[32] use the negative tangent modulus technique to unload the failed ply together with a maximum strain concept to uncouple lamina stiffnesses. An alternate approach due to Sandhu[33] is similar except it uses a maximum strain energy concept to uncouple the familed lamina.

It should be noted that interlamina shearing is not taken into account in these failure theories which sometimes represents a serious limitation.

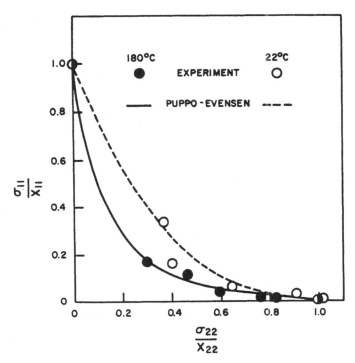

Fig. 8. Comparison of normalized experimental and predicted off-axis
         strengths (ramp loaded uniaxially to failure) at room tempera-
         ture (22°C) and glass-transition temperature (180°C).

Time Independent Fracture Theories

Composite laminates often contain numerous flaws due to the nature
of the manufacturing process.[34-35] Because of this fact, the application of linear elastic fracture mechanics (LEFM) to composite materials
has been extensively explored.

Waddoups, Eisenmann, and Kaminski[36] were among the first to apply
LEFM to notched composites. They studied the effect of centrally
located circular holes and cracks on the strength of uniaxial tension

strips. The solution for a crack emanating from a circular hole was
used for the tensile strip containing a hole even though no actual crack
existed. They argued that the material contained a damaged area adjacent
to the hole which was analogous to a crack. Mode I LEFM was used, and
the crack size was adjusted using an Irwin type plasticity correction
factor.[37] In a similar manner a damage region was assumed ahead of the
crack in the center cracked tensile strip. In essence, the usual one-
parameter LEFM model was replaced by a two-parameter LEFM model in which
the length of the damage zone was the second parameter.

Nuismer and Whitney[38] also adopted a two-parameter approach similar
to the inherent flaw model of Waddoups, et al. They likewise investigated tensile strips containing both centrally located circular holes and
cracks and introduced different fracture models called the point and
average stress criteria. In each case a characteristic size was employed
or a damage zone was assumed to exist immediately in front of the hole
or crack.

A material modeling approach to the fracture of notched laminates
has been proposed by Zweben[39] and Rosen, Kulkarni, and McLaughlin.[40]
In this approach the region around the notch tip was subdivided into
zones of differing types of behavior, resulting in an approximate model
designed to incorporate the major influences on the phonomenon under consideration. A predominate characteristic of the model was the ability to
account for other failure mechanisms as well as self-similar crack growth.

A local heterogeneous region (LHR) concept was developed by
Kanninen, Rybicki, and Griffith.[41] The approach treats the material as

heterogeneous and anisotropic where microstructural effects predominate and as homogeneous and anisotropic elsewhere. The model employed extensional and rotational spring-type elements in such a way that non-self-similar crack growth could be predicted. Cruse[42] related the fracture toughness of an arbitrary laminate to the fracture toughness of its individual ply constituents and basic laminate properties. One important conclusion relevant to obtaining valid experimental fracture toughness data was that crack length had to be large relative to damage zone size. Later Snyder and Cruse[43] presented a boundary-integral equation (BIE) method based on LEFM applied to homogeneous anisotropic materials. Their approach is valid for a variety of in-plane loading conditions, does not require artificial finite width correction factors as some other models and a characteristic dimension or damage zone is not needed.

The complications of anisotropy and heterogeneity aside, probably the most serious limitation to applying LEFM to composite is that self-similar crack propagation is a necessary condition for the theory to be applicable. For isotropic materials such as metals under mode I loading, self-similar growth is usual, but for composite laminates self-similar growth is the exception rather than the rule. Part of the reason for this is the nature of the material. A variety of fracture mechanisms such as fiber breakage, matrix cracking, fiber-matrix delamination, laminae delamination, bridging, etc., are factors in the ultimate fracture or separation of a laminate.

The point and average stress criteria of Whitney et al. and the inherent flaw model of Waddoups et al. can be expressed as, respectively,[36,38]

$$\frac{\sigma_n^\infty}{\sigma_u} \begin{cases} (1 - \xi)^{1/2} \\ [(1 - \xi)/(1 + \xi)]^{1/2} \\ [a_o/a + a_o]^{1/2} \end{cases} \qquad (31)$$

where σ_n^∞ is the ultimate tensile strength of an infinitely wide plate

with a center crack, σ_u is the unnotched laminate strength, $\xi = \dfrac{a_o}{a + a_o}$

and a is the half crack length. The parameter a_o is defined as the

characteristic length for each model. Further, $\sigma_n^\infty = Y\sigma_n$ where Y is a

finite width correction factor and σ_u is the ultimate unnotched strength

of the finite width test specimen.

The BIE method for determining mode I and mode II stress intensity

factors is an integral of boundary displacements and tractions on an

arbitrary path in the body, and is given by[43,44]

$$K_{I,II} = \int_s R_i^{I,II}(Q) u_i(Q) ds(Q) + \int_s L_i^{I,II}(Q) t_i(Q) ds(Q) \qquad (32)$$

The kernels R_i and L_i are complex variable functions which include crack

size and location, as well as material properties. The quantities

$u_i(Q)$ and $t_i(Q)$ refer to, respectively, displacements and tractions at

boundary points Q, exclusive of the crack boundary. The only input in-

formation needed in the technique is geometry, material properties, and

grid sizes and spacings.

A more detailed review of the above fracture theories may be found

in references 44-46. Figure 9 shows comparisons between experimental

data and several of the aforementioned theories applied to T300/934 G/E

laminates.

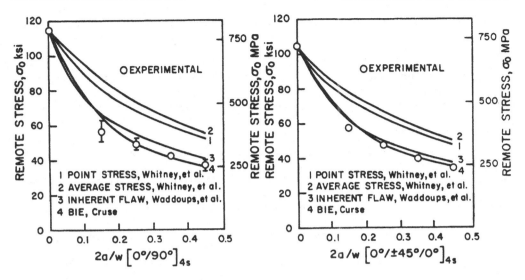

Fig. 9. Comparison between experimental and analytical critical remote
         stresses for T300/934 G/E laminates.

Comparisons between experimental and analytical remote critical
stresses were generally poor when using the two-parameter characteristic
models. However, the analytical predictions were made assuming a constant
characteristic length for all crack sizes and laminates. Better results
could have been obtained by using a "best fit" characteristic length, as
discussed in reference 47. On the other hand, the BIE method gave excellent predictions of experimental observations for aspect ratios greater
than 0.25.

The BIE method can be extended to other geometries and types of
loading, and does not use a characteristic length that may be dependent
upon material type and laminate orientation. By comparison, the two-
parameter models depend upon an analytical solution that is dependent
upon both the geometry and type of loading. Thus, the BIE method offers
an approach that is more general than the two-parameter approach, and

should be extended to include problems of different geometries and loadings.

Birefringent coatings can be used to visualize the intense energy
regions adjacent to cracks and flaws. Fig. 10 shows isochromatic
patterns for several laminates.[44,45]

In general, for $[\pm45°]_{4s}$ laminates, the fringe pattern tends to be
more dense in the +45° direction than the -45° direction adjacent to the
hole. As the outer ply was in the +45° in each case, this observation
merely indicated that the outer ply was constrained by the inner ply to
give this effect. The intense stress region for $[0°/90°]_{4s}$ laminates is
in the direction of the load and/or the 0° fibers.

The isochromatic patterns for the $[0°/45°/0°]_{2s}$ laminates are
generally similar to those in isotropic materials, i.e., note the butter-
fly wing shaped pattern in Fig. 10. The direction of initial notch tip
fracture tended to correspond to the angle of inclination of the fringes
at the notch tip. Caution must be used in the quantification or
birefringence results because as pointed out by Dally and Alfirevich,[48]
the mismatch between the properties of the coating and the underlying
material (Poisson's ratios and moduli) as well as the anisotropy of the
laminate creates difficulties for proper fringe interpreation. Also, the
nature of the singularity in the laminate may be quite different than the
nature of the singularity in the photoelastic coating.

Linear Viscoelastic Effects

The isothermal linear anisotropic theory of viscoelasticity is
completely defined by the relaxation modulus $Q_{ijk\ell}(t)$, and the creep

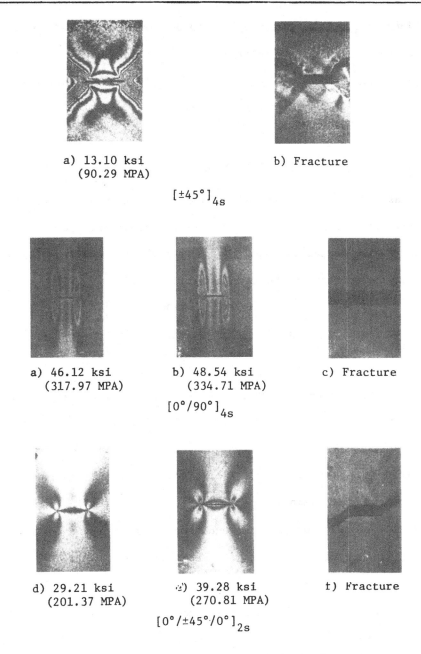

a) 13.10 ksi   b) Fracture
   (90.29 MPA)

$[\pm 45°]_{4s}$

a) 46.12 ksi   b) 48.54 ksi   c) Fracture
   (317.97 MPA)   (334.71 MPA)

$[0°/90°]_{4s}$

d) 29.21 ksi   e) 39.28 ksi   f) Fracture
   (201.37 MPA)   (270.81 MPA)

$[0°/\pm 45°/0°]_{2s}$

Fig. 10. Birefringent coatings on T300/934 G/E laminate.

compliance, $S_{ijk\ell}(t)$. The response to any arbitrary strain or stress input may be found by superposition using the convolution integral which may be written as

$$\sigma_{ij}(t) = \int_0^t Q_{ijk\ell}(t - \tau) \frac{d\varepsilon_{k\ell}}{d\tau} d\tau \tag{33}$$

and

$$\varepsilon_{ij}(t) = \int_0^t S_{ijk\ell}(t - \tau) \frac{d\sigma_{k\ell}}{d\tau} d\tau \tag{34}$$

Equations (33) and (34) are the viscoelastic equivalent of equations (1) and (2). For a lamina in a state of plane stress only four time dependent properties are needed analogous to those given in equations (4) and (5). These may be written in the matrix form,

$$S(t) = \begin{bmatrix} S_{11}(t) & S_{12}(t) & 0 \\ S_{21}(t) & S_{22}(t) & 0 \\ 0 & 0 & S_{66}(t) \end{bmatrix} \tag{35}$$

As with the linear elastic case, the linear viscoelastic compliance tensor is symmetric. Schapery[49] has verified this analytically as long as each of the constituent phases is symmetric. Yeow, et al.[26,50] have borne this out experimentally. Also, Yeow, et al.[26,51] showed that a unidirectional G/E laminate loaded in the direction of the fibers was essentially elastic for the time and temperature ranges considered, i.e., $S_{11}(t)$ = constant and $S_{12}(t)$ = constant.

For composite laminates the creep compliance is experimentally easier to measure than the relaxation modulus. In a creep test the strain may be measured at the center of the specimen, the stress calculated from the applied load, and the compliance calculated directly. When the creep specimen is of adequate length, end effects negligibly perturb the stress at the center and an accurate calculation of compliance can be made. In a relaxation test several problems present themselves. First, any specimen movement in the grips or less than adequate testing machine stiffness will artificially change the relaxation function. Secondly, in a relaxation test the strain is usually measured at the center using a strain gage or extensometer. When the relaxation modulus is calculated, it is implicitly assumed that the measured strain is valid for the length of the specimen and that the specimen is free of constraints so a uniaxial stress field is present. This condition of course cannot be satisfied at the ends due to the constraint of the grips, or lack of grip rotation. Further, when a ply fails or ruptures other plies can creep even though the overall specimen is in a fixed grip situation.[9,52]

In a unidirectional laminate necessary values of compliance may be determined from three uniaxial tensile creep tests similar to those previously discussed. These may be specified as,

1. For a uniaxial test of a 0° specimen

$$S_{11}(t) = \frac{\varepsilon_1(t)}{\sigma_1^o} \quad \text{and} \quad S_{21}(t) = \frac{\varepsilon_2(t)}{\sigma_1^o} \tag{36}$$

2. For a uniaxial test of a 90° specimen

$$S_{22}(t) = \frac{\varepsilon_2(t)}{\sigma_2^o} \quad \text{and} \quad S_{12}(t) = \frac{\varepsilon_1(t)}{\sigma_2^o} \tag{37}$$

3. For a uniaxial test of any orientation not 0° or 90° one may use the transformation equation and the results of tests 1 and 2 to calculate S_{66} using the measured off-axis compliance, S_{xx} from the following,

$$S_{xx}(t) = \cos^4\theta \; S_{11}(t) + \sin^4(\theta) \; S_{22}(t)$$
$$+ \cos^2\theta \; \sin^2\theta \; (2S_{12}(t) + S_{66}(t)) \qquad (38)$$

Alternatively, when the complete state of strain is known

via a strain gage rosette S_{66} may be calculated using

$$S_{66}(t) = \frac{\gamma_{12}(t)}{\tau_{12}^o} \qquad (39)$$

In the above the superscript for stress has been added to indicate the

value of stress which is applied at time equal zero and held constant

thereafter. As discussed previously, our preference is the 10° off-axis

tensile test.

Fig. 11 shows the results of a creep test for a G/E material at

200°C.[26]

Time-Temperature Superposition Principle

The time-temperature superposition principle (TTSP) is often

referred to by a number of different names including Time Temperature

Analogy, Method of Reduced Variables, Time Translation Equivalence and

others. The TTSP was first developed for polymers and has only been

recently applied to composites by Brinson and his colleagues.[9,51]

The TTSP provides a method by which long-term behavior can be pre-

dicted from short-term tests (accelerated characterization). Initially

the technique was applied to the relaxation modulus and the creep

compliance. Subsequently, it has been applied to a number of other

properties including failure.[54]

The history of TTSP has been recently summarized in detail by

Markovitz.[55] In his description he points out that the knowledge of the

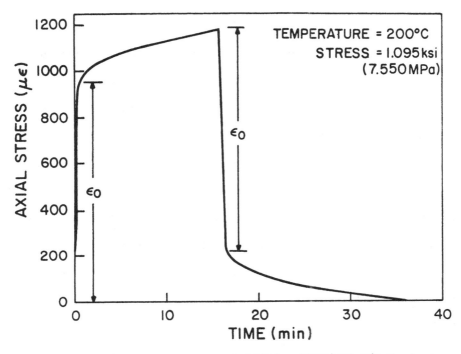

Fig. 11.   Creep and creep recovery of $[90°]_{8s}$ T300/934 G/E laminate.

general effects of temperature on creep and relaxation were known by a

number of investigators in Russia, Europe and the United States.

Markovitz, however, finds no explicit reference to TTSP prior to

Leaderman's 1943 report.   The basic idea is that compliance curves at

different temperatures are of the same basic shape, but only shifted in

time.   Thus by taking short term compliance data at several temperatures

and then shifting these curves horizontally in log time one can obtain a

smooth curve approximating the compliance over many decades of time.

For Thermorheologically Simple Materials (TSM), i.e., those for

which only a horizontal shift is needed, the compliance for an arbitrary

temperature T is given by,

$$D(T,t) = D_o(T_o) + \int_{-\infty}^{\infty} L(T_o, \ln \tau)[1 - e^{-\xi/\tau}]d \ln \tau \tag{40}$$

where T_o is a reference temperature, ξ is the reduced time given by,

$$\xi = t/a_T \tag{41}$$

and a_T is the temperature shift factor. When the retardation spectrum and compliance are plotted vs a log time scale, the effect of a_T is merely to shift these curves to the right or left in time according to,

$$\log \xi = \log t - \log a_T \tag{42}$$

Unfortunately, most engineering materials do not fit into the TSM description, and are classified as thermorheologically complex materials (TCM). For this case, there will be a vertical shift in the retardation spectrum and compliance, as well as a horizontal shift. The compliance at a temperature T is given by,

$$D(T,t) = D_o(T) + \int_{-\infty}^{\infty} L(T, \ln \tau)[1 - e^{-\xi/\tau}]d \ln \tau \tag{43}$$

For many materials and temperature ranges, however, the temperature dependence of D_o and L tends to be fairly small. The horizontal shift for various temperatures remains the fundamental concept.

To use the TTSP for either TSM or TCM, compliance data is taken for a number of different temperatures. The duration of these tests is normally quite short because of practical considerations. This short term data is then shifted horizontally and vertically to form a smooth and continuous "master curve" which is assumed to be valid over many decades of time at an arbitrary reference temperature. To obtain the compliance at other temperatures, the master curve is shifted to coincide with the short term data at that temperature.

Graphical shifting and/or normalization procedures have been described in detail by Griffith.[25] These range from the simple Tobolsky-Ferry vertical shift procedure to the much more accurate and complete vertical shift procedure of McCrum and Morris.[57] Fundamentally, the former is valid only above the glass-transition temperature, T_g, while the latter is valid over the complete range of polymeric response from glassy to rubbery. Griffith's main point was that all shifting can be done graphically without need of a formal vertical or horizontal shift relationship.

Yeow[26] measured the T_g for a unidirectional G/E laminate and applied the TTSP technique to tensile test data for numerous off-axis uni-directional specimens. The thermal expansion of a $[90°]_{8s}$ laminate is shown in Fig. 12 from which the T_g was estimated. Fig. 13 shows the basic data needed for a $[10°]_{8s}$ laminate to obtain the reciprocal of compliance, $1/S_{xx}(t)$, master curve shown in Fig. 14. For the data shown in Figs. 13 and 14, a Tobolsky-Ferry vertical shift procedure was used.

Also, shown in Fig. 14 are the results of an orthotropic transforma-tion given earlier as equation (38). Good correlation was found between the master curve based on short time (15 minute) data, the transformation equation calculations based on long time master curves and the results of long term tests (10^3 minutes).[9,26,51]

Time-Temperature-Stress Superposition Principle (TTSSP)

Efficient utilization of composites almost certainly requires structural components to be highly stressed. For this reason, methods are needed to account for material nonlinearities created by high stress

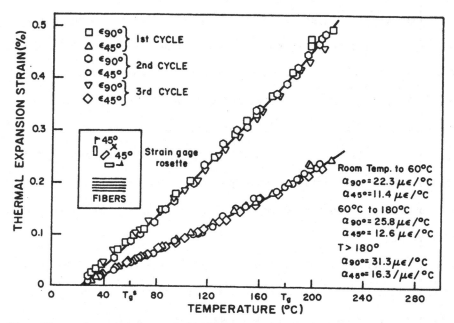

Fig. 12. Thermal expansion of [90°]$_{8s}$ T300/934 G/E laminate with glass-transition temperatures indicated.

levels. The time-temperature-stress superposition principle is one such method.

In essence, the time-temperature-stress superposition principle is a simultaneous application of the previously discussed TTSP and an analogous time-stress superposition principle (TSSP). In the former, an increase in temperature is assumed to accelerate a sequence of deformation events and in the latter an increase in stress is assumed to accelerate a sequence of deformation events. Mechanisms, of course, are assumed to remain unchanged in both cases. The combined TTSSP was first used by Daugste[58] to predict the nonlinear viscoelastic behavior of a 45° glass reinforced unidirectional composite.

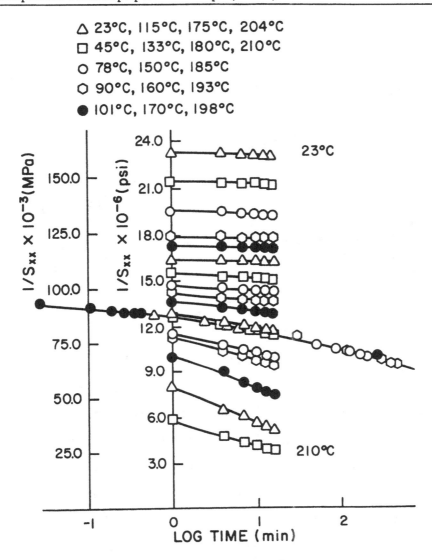

Fig. 13.   Reduced reciprocal of compliance, $1/S_{xx}$, and portion of 180°C master curve for $[10°]_{8s}$ T300/934 G/E laminate.

The effects of moisture can be studied through an analogous time-temperature-stress-moisture superposition principle. Crossman et al.[59] have explored the use of a time-temperature-moisture superposition principle.

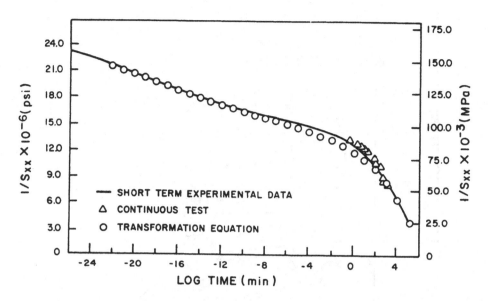

Fig. 14. Master curve of the reciprocal of reduced compliance, $1/S_{xx}$, of $[10°]_{8s}$ laminate at 180°C.

For illustrative purposes hypothetical transient creep compliance vs. log time is shown in Fig. 15 for several stress and temperature levels. The data from Fig. 15a for each temperature level may be shifted to obtain the σ_1 master curve shown in Fig. 15b using the TTSP. Similarly, master curves may be formed for stress levels σ_2, σ_3 and σ_4. An outcome of this procedure will be the temperature shift factor, log a_T, and its corresponding stress dependence. The data from Fig. 15c for each stress level may be shifted to obtain the T_1 master curve shown in Fig. 15d using the TSSP. Similarly, master curves may be formed for temperature levels of T_2, T_3 and T_4. This procedure will yield the stress shift factor, log a_σ, and its associated temperature dependence. The master

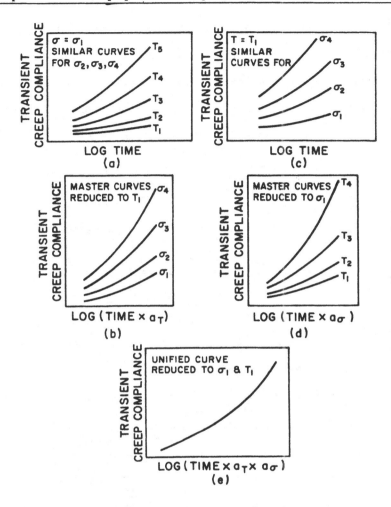

Fig. 15.   Schematic diagram to illustrate the time-stress-temperature
          superposition principle.

curves in Fig. 15b or the master curves in Fig. 15d may now be shifted to

obtain the unified master curve for a stress σ_1, and a temperature T_1, as

shown in Fig. 15e.   This unified master curve can now be shifted to

determine a unified master curve for any temperature and/or stress level

within the range of data.

Yeow[26] and Griffith[25] extensively reviewed the literature related
to the TTSSP method. Griffith further applied the technique to a particular G/E laminate. He produced S_{22} and $S_{10°}$ creep compliance master
curves. Individual $S_{10°}$ creep compliance curves for 290°F (143°C), 320°F
(160°C), 350°F (177°C) and 380°F (193°C) are given in Figs. 16-19,

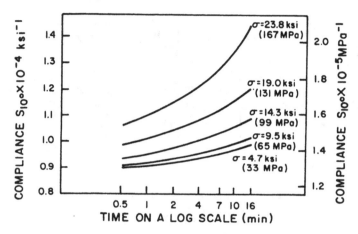

Fig. 16. $S_{10°}$ compliance at 290°F (143°C) as a function of stress
         level for T300/934 G/E laminate.

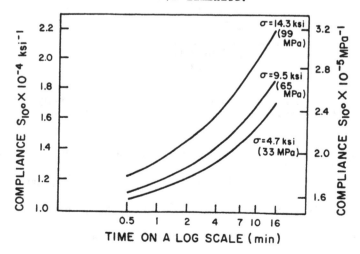

Fig. 17. $S_{10°}$ compliance at 320°F (160°C) as a function of stress level
         for T300/934 G/E laminate.

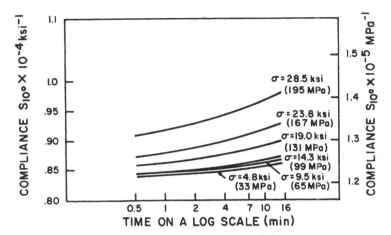

Fig. 18. $S_{10°}$ compliance at 350°F (177°C) as a function of stress level for T300/934 G/E laminate.

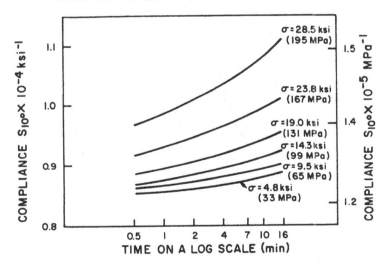

Fig. 19. $S_{10°}$ compliance at 380°F (193°C) as a function of stress level for T300/934 G/E laminate.

respectively. The resulting master curve is shown in Fig. 20 and the associated temperature and stress dependent shift function surface is shown in Fig. 21. Also shown in Fig. 20 is a comparison between the $S_{10°}$ master curve and the results of a long-term creep test in excess of

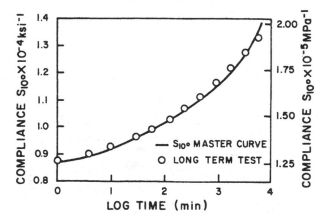

Fig. 20.   Comparison of $S_{10°}$ master curve for T300/934 G/E laminate with a long term test at 320°F (160°C) and σ = 19,500 psi.

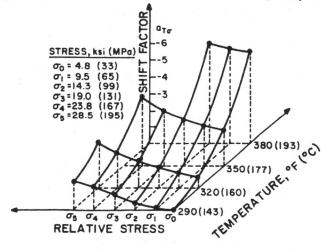

Fig. 21.   Shift surface for combined shift factor, $a_{T\sigma}$, for $S_{10°}$ for T300/934 G/E laminate.

150 hours. It appears that at extremely long times the master curves may tend to over-predict the compliance. The reason for this may be due to additional curing or more likely some other form of aging of the material. Nevertheless, the agreement between predictions and experiment is reasonable. More details about postcuring and aging may be found in

references 60 and 61.

An S_{66} master curve was generated using equation (38). Further, the S_{22} and S_{66} stress dependent master curves were used in conjunction with the transformation equation to predict the long term compliance of $[30°]_{8s}$ and $[60°]_{8s}$ specimens at 320°F (160°C). Predictions for the $[60°]_{8s}$ laminate are compared with long term test results in Fig. 22. Agreement is seen to be fair to good.

Schapery Approach to Nonlinear Viscoelasticity

Another nonlinear approach of interest is that proposed by Schapery.[62] His approach is derived from thermodynamic considerations and has been used successfully by several investigators[63-66] to predict

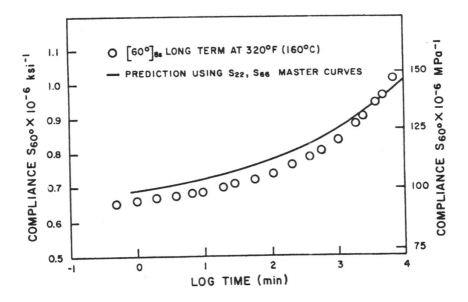

Fig. 22. Comparison of predicted and measured $S_{60°}$ compliance at 320°F (160°C).

the behavior of polymers both with and without fiber reinforcement. The
form of the constitutive equation for uniaxial stress is given by,

$$\varepsilon(t) = g_o D_o + g_1 \int_{-\infty}^{t} \Delta D(\psi - \psi') \frac{dg_2 \sigma}{d\tau} d\tau \qquad (44)$$

where g_o, g_1, and g_2 are functions of the streys level, D_o is the
instantaneous component of the linear viscoelastic compliance, ΔD is the
transient component of the linear viscoelastic compliance, ψ and ψ' are
reduced time parameters as given by,

$$\psi = \psi(t) = \int_o^t \frac{dt'}{a_\sigma} \qquad ; \qquad \psi' = \psi'(\tau) = \int_o^\tau \frac{dt'}{a_\sigma} \qquad (45)$$

where a_σ is the stress dependent time shift factor. The basic form is
very similar to the Boltzman superposition integral and, in fact, in the
linear range of the material when σ is small, $g_o = g_1 = g_2 = a_\sigma = 1$, the
Boltzman integral is regained. Furthermore, the reduced time and shift
factor concept is also employed. It seems reasonable to hypothesize ex-
tending this procedure to include the features of temperature superposition by perhaps letting,[25]

$$\varepsilon(t) = g_o(\sigma) D_o(T)\sigma + g_1(\sigma) \int_{-\infty}^{t} \Delta D(T, \psi-\psi') \frac{d(g_2(\sigma)\sigma)}{d\tau} d\tau \qquad (46)$$

where ψ and ψ' are now reduced times with respect to both stress and
temperature shift factors,

$$\psi = \psi(t) = \int_o^t \frac{dt'}{a_\sigma a_T} \qquad ; \qquad \psi' = \psi'(\tau) = \int_o^\tau \frac{dt'}{a_\sigma a_T} \qquad (47)$$

While this approach has not been pursued as yet, the development of a
unified technique to account for both temperature and nonlinear stress
effects would be very advantageous.

The Schapery procedure is very appealing from the standpoint that it provides a unified approach to predicting the nonlinear viscoelastic response to an arbitrarily varying stress. However, difficulties do arise in the experimental determination of g_0, g_1, g_2, and a_σ. Because the approach is more general, more information is required to evaluate the above stress dependent functions. In particular, Schapery uses creep and creep recovery data to determine the unknown functions. He also proposed that the transient compliance be modeled by a simple power law which is not a function of stress.

$$\Delta D(\psi) = m \ \psi^n \tag{48}$$

In the preceding approach, the necessity for a stress dependent shift factor, a_σ, is created. This shift factor is analogous to the temperature shift factor, a_T, and should be the same as the stress dependent shift factor obtained by application of the time-stress superposition principle (TSSP). The two shift factors a_T and a_σ can be combined to give a single shift factor $a_{T\sigma}$ which should be the same as the shift factor in the TTSSP method. That is, the shift factor in equations (47) should be the same as the shift factor of Fig. 21 for the same material.

Cartner and Brinson[66] applied Schapery's procedure to a chopped fiber glass SMC material. Creep recovery curves as a function of stress level are shown in Fig. 23. These curves were shifted horizontally and vertically to obtain the master recovery curve shown in Fig. 24.

Fig. 25 shows the comparison between measured creep strains and those predicted by the Schapery procedure.

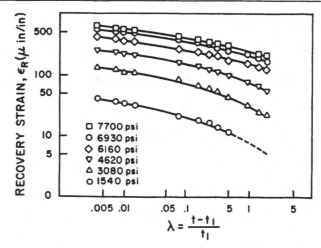

Fig. 23.   Recovery strains for various stress levels for SMC at 25°C.

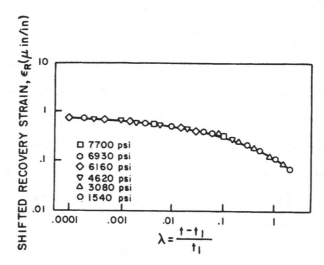

Fig. 24.   Master recovery curve for SMC at 25°C.

Time Dependent Failure Theories

A possible mathematical model for the characterization of a uniaxial

delayed yield phonomenon was proposed by Crochet.[67] He assumed the yield

stress to be a monotonically increasing function for increasing strain

rate which implies that a faster loading rate produces a higher yield

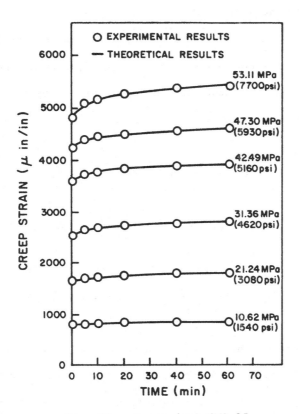

Fig. 25.   Creep data SMC-25.

stress.   The yield stress was given the specific form such that,

$$\sigma_f(t) = A + B \exp(-D\chi) \qquad (49)$$

where $\sigma_f(t)$ is a time dependent yield stress and A, B and D are material

constants.   The time dependent material function, χ, was defined by

Crochet as

$$\chi = [(\epsilon_{ij}^{V} - \epsilon_{ij}^{E})(\epsilon_{ij}^{V} - \epsilon_{ij}^{E})]^{1/2} \qquad (50)$$

where ϵ_{ij}^{V} and ϵ_{ij}^{E} refer to the viscoelastic and elastic strains re-

spectively.   Equation (50) is a specialization of a more general theory

due to Nagdi and Murch.[68]

Brinson[69] used equations (49) and (50) in conjunction with a linear
modified Bingham viscoelastic-plastic model to accurately predict
delayed failures in polycarbonate. More recently, Cartner and
Brinson[65,66] used eqs. (49) and (50) in conjunction with Schapery's non-
linear theory to obtain an equation of the form

$$t_f = \left[\frac{1}{D\beta\sigma_f} \ \ell n \ \frac{\sigma_f - A}{B} \right]^{\frac{1}{n}} \qquad (a)$$

$$\beta = m \ \frac{g_1 g_2}{a_\sigma{}^n} \ (1 + 2\nu^2)^{1/2} \qquad (b) \tag{52}$$

where t_f is the creep time to failure or rupture at a stress σ_f, ν is
Poisson's ratio (assumed to be a constant) and all other quantities are
as earlier defined.

A unique feature of the above approach is that viscoelastic constitutive and fracture processes are combined. That is using this procedure
allows the prediction of stress-strain response from zero load to rupture
with an integrated analytical model. Such is not the case for the more
nearly empirical models subsequently discussed.

Fig. 26 shows the comparisons between measured creep rupture data[70]
and the predictions given by equation (52) for an SMC material.[65,66]

Several other creep rupture criteria for homogeneous isotropic
materials are based on a linearly decreasing logarithm of the time to
rupture with increasing stress. This form, as exemplified by the Zhurkov,
Larson-Miller, and Dorn methods, is given by

$$\log t_f = A - B\sigma \tag{53}$$

where t_f is the time to failure for a constant creep load of σ. A and B

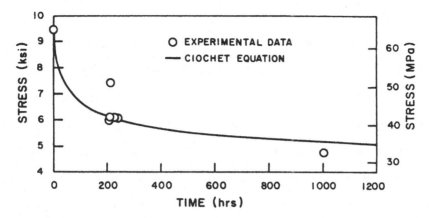

Fig. 26. Comparison of creep to rupture data with Crochet equation.

are material constants for a given temperature.[71] Landel and Fedors[72]

have noted that in some circles, the form,

$$\log t_f = A - B \log \sigma \tag{54}$$

is viewed more favorably.

Experimentally, the creep stress level is the independent variable
and the time to rupture at that stress level is the dependent variable.
For the analysis, however, it is convenient to rearrange equation (52)
to express the creep failure strength, σ_f, as a function of the time to
rupture,

$$\sigma_f = (A - \log t_f)/B . \tag{55}$$

Of the numerous orthotropic static failure theories available,[11,73]
the Tsai-Hill criteria was chosen by Dillard[35] to account for viscoelastic
effects. He modified the Tsai-Hill criteria to account for time
dependent creep rupture strengths as follows,

$$\frac{\sigma_1^2}{[X(t_f)]^2} + \frac{\sigma_1 \sigma_2}{[X(t_f)]^2} + \frac{\sigma_2^2}{[Y(t_f)]^2} + \frac{\tau_{12}^2}{[S(t_f)]^2} = 1 \tag{56}$$

Here, time independent strengths have been replaced by creep rupture strengths which result in failure at $t = t_f$ as defined by equation (55). $X(t_f)$ represents the creep rupture strength for a uniaxial creep load parallel to the fiber direction. The assumption was made that delayed failures do not occur for $0°$ specimens and that $X(t_f) = X$. $Y(t_f)$ represents the functional relation with time of the creep rupture strength for a uniaxial creep load perpendicular to the fiber direction. $S(t_f)$ is a similar shear creep rupture strength. Theoretically, $S(t_f)$ can be determined from uniaxial creep rupture of off-axis specimens and prior knowledge of X and $Y(t_r)$. Dillard indicated that such a procedure, though straightforward, proved unsatisfactory and, as a result, the shear creep rupture strength was assumed to be of the form

$$S(t_f) = \alpha \, Y(t_f) \ . \qquad\qquad (57)$$

To determine the value of α in equation (57), Dillard[35] used Griffith's[25] creep rupture data which is shown plotted in Fig. 27. Also, shown in Fig. 27 are the results of creep rupture tests on a postcured $[60°]_{8s}$ laminate. Basically little difference between unpostcured and postcured material is indicated for the time range for the investigation. Nevertheless, adequate postcuring of specimens prior to testing is highly recommended.[25,35]

The data of Fig. 27 is shown replotted in Fig. 28 together with the normalized time dependent Tsai-Hill criteria of equation (56). The information shown in Fig. 28 resulted in a value of $\alpha \cong 0.65$.[35]

Several other time dependent failure theories are worthy of note. A classic deformational approach is given by Landel and Fedors.[72] They

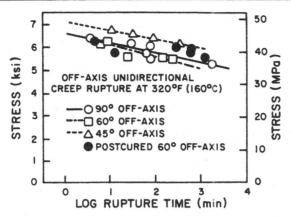

Fig. 27.   Creep rupture of off-axis T300/934 G/E unidirectional
           specimens at 320°F (160°C).

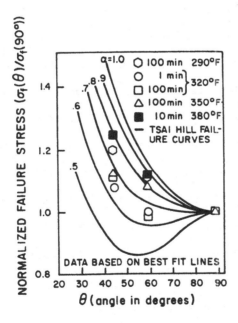

Fig. 28.   Normalized creep rupture versus fiber angle with parametric
           Tsai-Hill curves for a T300/934 G/E laminate.

develop a property surface in stress-strain-time space to establish the

constitutive behavior for several polymers. Furthermore, they suggest

that failure merely constitutes a boundary to the surface. Thus, the
property surface uniquely establishes the stress-strain-time relationship for any load path, and the surface will also give the stress and
strain at the time of failure. Their visualization was only for the case
of uniaxial stress or pure shear for a polymer.

Bruller[74] has established a creep energy criterion for correlating
the onset of crazing and the limit of linear viscoelasticity and has
applied the technique to polymethylmethacralate polycarbonate and an
epoxy.

Brinson[9] assumed the shape of the creep rupture strength master
curve has the same as the reciprocal of linear viscoelastic compliance
master curve. Different temperatures can be accommodated by merely
translating the strength master curve in accordance with the same shift
factors used for compliance. Thus, for a unidirectional composite,
strength master curves at an arbitrary temperature and fiber angle can
be found from the corresponding modulus or compliance curve and knowledge
of one generic strength data point by simple scaling procedures. Fig. 29
shows the results of the application of this process to the prediction of
creep rupture data for a $[60°]_{8s}$ T300/934 G/E laminate.[75] Good correlation with experiment was obtained. It should be noted that the form of
the predictions and data so obtained agree well with the form of equations (53) and (54).

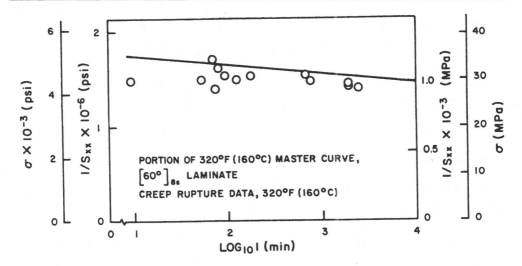

·Fig. 29.   Comparison of T300/934 creep rupture predictions and experimental results. Predictions assume strength and modulus
master curves have the same shape.

Accelerated Characterization of Laminate Compliance and Creep Rupture

Predictions

The preceding sections discuss many of the details necessary to the

utilization of the accelerated characterization plan shown in Fig. 1 and

presented earlier. Initial efforts to verify the plan were based upon

elements of linear viscoelasticity for small stresses and strains and

the TTSP. The generic ramp loaded properties (item A) and modulus

transformations were presented in Fig. 6.[10] Time and temperature

dependent lamina strength predictions (items C and G) were made using a

modified Puppo-Evenson failure criteria, an illustration of which was

given in Fig. 8.[26]

Examples of TTSP master curves (items D and F) for small stresses

and strains were given in Figs. 13 and 14.[9,26,51] A Tobolsky-Ferry type of vertical shift was used for all data. In retrospect, the use of this type of vertical shift below the T_g is likely incorrect.[25] The shift function relationship (item E) was shown to conform to the WLF equation[51] above the T_g, but to be generally nonlinearly related to the reciprocal of absolute temperature.[26] The latter implies that a linear Arrhenius equation is inapplicable. Shift functions were further shown to be essentially independent of fiber angle (item E) but only for small stresses and strains.[26,50]

As reported by Griffith,[25] Yeow developed a linearly viscoelastic incremental laminated theory (item H) and predicted the creep ruptures for $[\pm 45°]_{4s}$ and $[90°/\pm 60°/90°]_{2s}$ laminates. Griffith performed numerous creep rupture tests (item I) and compared his results with predictions. These comparisons are shown in Fig. 30 and, as may be observed, substantial deviation between measurement and analysis was found. The poor correlation is not surprising as the analytical predictions were based on linear viscoelastic behavior which was known to be valid only for small stresses and strains.

As a result of the poor correlation shown in Fig. 30, Griffith[25] performed an exhaustive study using TTSSP principles which identified the stress and temperature dependent non-linear viscoelastic nature of the same material. He further studied various time dependent failure theories and compared these with measured results. Examples of Griffith's work were shown in Figs. 16-22 and 27.

Dillard,[35] using Griffith's data and some of his own, developed a non-linearly viscoelastic incremental lamination theory for predicting

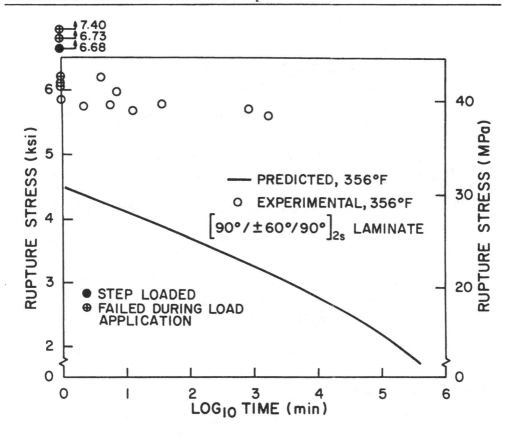

Fig. 30. Comparison of $[90°/±60°/90°]_{2s}$ creep rupture predictions and experimental results for T300/934 G/E laminates.

the degradation with time of laminate compliances and strengths (items H and I of Fig. 1). The analytical model contained numerous features including a cumulative damage concept, stress interaction model, a modified Tsai-Hill-Zhurkov failure theory (Fig. 28), non-linear super-position principle, first ply failure and unloading fractures, etc. An example of the predictive capability of his analytical model for a $[90°/±60°/90°]_{2s}$ laminate is shown in Fig. 31. Reasonable correlation was found for the time scale of the measurements.[35,76]

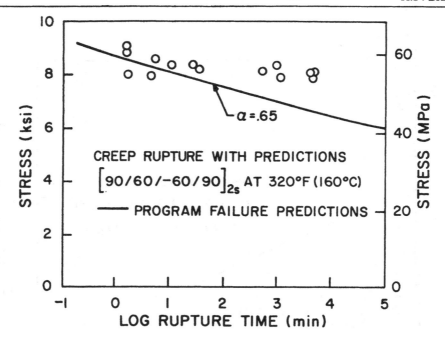

Fig. 31. Creep rupture data with predictions for T300/934 G/E laminate.

Summary and Conclusions

The objective herein was to describe the application of methods of experimental mechanics to the accelerated characterization of visco-elastic processes in polymeric composites including delayed failures. Elements of an accelerated characterization plan was presented. The essential idea for the plan was to develop a procedure by which long time properties could be predicted from short time test results, perhaps no more extensive than those needed for routine quality control procedures. The technique outlined was based upon a master curve concept for lamina modulus and strength properties and an incremental viscoelastic lamina-tion theory. The measurement of lamina modulus and strength properties were discussed. Also, the analytical characterization of time independent and dependent properties was discussed in detail using several different methods.

Time independent fracture theories were discussed as was the application of birefringent coatings to visualize regions of high stress gradients around cracks. Time dependent fracture theories were not discussed herein nor are there generally acceptable models available for composite laminates. Obviously, the eventual prediction of laminate creep rupture processes will, of necessity, need to include fracture mechanics concepts for viscoelastic materials.

The integration of measured lamina properties and their analytical characterization to predict laminate creep ruptures was briefly presented.

There is no question, at least in the author's mind, that matrix dominated properties of polymer based composite materials are viscoelastic and are important to their long term integrity. Further, there is no question that delayed failures do occur in both chopped fiber SMC materials and G/E laminates. The safe design of properties utilizing these structural materials will require knowledgeof the kinds of elements attempted to be studied and discussed herein.

Perhaps in conclusion a word is in order relative to the nature of experimental mechanics in general and as applied to composite materials in particular. One can take the view, and many do, that experimental mechanics is in fact the knowledgeable use of various and sundry methods and apparatuses such as extensometers, testing machines, photoelasticity, moiré, holography, etc. The author's view, however, is that experimental mechanics is in reality the knowledgeable application of the principles of mechanics to understand and characterize phenomena observed in the laboratory or in practical applications. Thus one cannot do experimental mechanics without knowledge of the fundamental laws of mechanics. Sometimes, of course, our observations lead to the development of, or modification of, fundamental laws. In these respects, composite materials offer a unique opportunity to use, understand and to better define the underlying principles of experimental mechanics.

Acknowledgements

The author is deeply indebted to several organizations for their financial support over a number of years which has assisted in the production of the current effort. Notably among these are NASA-Ames, NASA-Langley and the General Motors Corp. Quite a number of people have given their encouragement as well. Some of these include Dr. Howard G. Nelson, NASA-Ames; Mr. Dell P. Williams, NASA-Headquarters; Dr. M. F. Kanninen, Battelle Columbus; Dr. John G. Davis, NASA-Langley; Dr. Roger A. Heimbuch and Ms. Barbara Sanders, General Motors and many of my V.P.I. and S.U. colleagues. The author would, of course, like to acknowledge the outstanding work of his M.S. and Ph.D. students who have contributed to the present topic over the years. These include Drs. Y. T. Yeow, W. I. Griffith, and D. A. Dillard; Mr. M. S. Cartner, Mr. D. G. O'Connor, Mr. T. S. Massa, Mr. J. Mensch and Mr. A. Bertolotti. Appreciation is also due to Dr. J. Pindera for his constant encouragement and perseverance.

References

1. Holister, G. S. and Thomas C., Fibre Reinforced Materials, Elsevier, NY, 1966.

2. Calcote, L. R., The Analysis of Laminated Composite Structures, Van Nostrand Reinhold, NY, 1969.

3. Tsai, S. W., Halpin, J. C. and Pagano, N. J., Composite Materials Workshop, Technomic, 1968.

4. Ashton, J. E. and Whitney, J. M., Theory of Laminated Plates – Progress in Materials Science, Vol. 4, Technomic, Stanford, CT, 1970.

5. Broutman, L. J. and Krock, R. H., (eds.), Composite Materials, Vols. 1-8, Academic Press, NY, 1974.

6. Jones, R. M., Mechanics of Composite Materials, McGraw-Hill, NY, 1975.

7. Tsai, S. W. and Hahn, H. T., Introduction to Composite Materials, Technomic, Westport, 1980.

8. Lekhnitskii, G. S., Theory of Elasticity of an Anisotropic Elastic Body, Holden-Day, San Francisco, 1963.

9. Brinson, H. F., Morris, D. H., and Yeow, Y. T., "A New Experimental Method for the Accelerated Characterization and Prediction of the Failure of Polymer-Based Composite Laminates," 6th International Conference for Experimental Stress Analysis, Munich, West Germany, Sept. 1978. Also, VPI-E-78-3, Feb. 1978.

10. Yeow, Y. T. and Brinson, H. F., "A Comparison of Simple Shear Characterization Method for Composite Laminates," Composites, Jan. 1978, pp. 49-55.

11. Sandhu, R. S., "A Survey of Failure Theories of Isotropic and Anisotropic Materials," Tech. Rep. AFFDL-TR-72-71, Air Force Flight Dynamics Lab., Wright-Patterson Air Force Base, OH.

12. Bert, C. W., "Static Testing Techniques for Filament-wound Composite Materials," Composites, January, 1974.

13. Chamis, C. C. and Sinclair, J. H., "Ten-Degree Off-Axis Test for Shear Properties of Fiber Composites," Experimental Mechanics, Sept. 1977, pp. 354-358.

14. Daniels, B. K., Harakas, H. K., and Jackson, R. C., "Short Beam Shear Tests of Graphite Fiber Composites," Fiber Science Technology, March, 1971.

15. Cooper, G. A. and Kelly, A., "Tensile Properties of Fiber-Reinforced Materials: Fracture Mechanics," J. Mech. Phys. Solids, Vol. 15, 1967, pp. 279-297.

16. Durchlaub, E. C. and Freeman, R. B., "Design Data for Composite Structure Safe-life Predictions," AFML-TR-73-225, March 1974.

17. Pagano, N. J. and Halpin, J. C., "Influence of End Constraints in the Testing of Anisotropic Bodies," J. Composite Materials, Vol. 2, January 1968, pp. 18-31.

18. Petit, P. H., "A Simplified Method of Determining the In-plane Shear Stress-Strain Response of Unidirectional Composites," ASTM STP 460, 1969, pp. 83-93.

19. Rosen, B. W., "A Simple Procedure for Experimental Determination of the Longitudinal Shear Modulus of Unidirectional Composites," J. Composite Materials, Vol. 6, October 1972, pp. 552-554.

20. Sims, D. F., "In-Plane Shear Stress-Strain Response of Uni-directional Composite Materials," J. Composite Materials, Vol. 7, January 1973, pp. 124-128.

21. Whitney, J. M., Stansbarger, D. L., and Howell, H. B., "Analysis of the Rail Shear Test--Applications and Limitations," J. Composite Materials, Vol. 5, January, 1971, pp. 24-34.

22. Cole, B. W. and Pipes, R. B., "Filamentary Composite Laminates Subjected to Biaxial Stress Fields," Technical Report AFFDL-TR-73-115, Air Force Flight Dynamics Laboratory, Wright-Patterson Air Force Base, Ohio, June 1974.

23. Ashton, J. E. and Love, T. S., "Shear Stability of Laminated Anisotropic Plates," Testing and Design, ASTM STP 460, 1969, pp. 352-361.

24. Browning, C. E., Husman, G. E., and Whitney, J. M., "Moisture Effects in Epoxy Resin Matrix Composites," ASTM-STP 617, American Society for Testing and Materials, 1977, pp. 481-496.

25. Griffith, W. I., "The Accelerated Characterization of Viscoelastic Composite Materials," Ph.D. Thesis, May 1979. Also, VPI-E-80-15, April 1980.

26. Yeow, Y. T., "The Time-Temperature Behavior of Graphite/Epoxy Laminates," Ph.D. Thesis, VPI&SU, May 1978.

27. Ashkenazi, E. K., "Problems of the Anisotropy of Strength," Mekhanika Polimerov, Vol. 1, No. 2, 1965.

28. Hill, R., "A Theory of the Yielding and Plastic Flow of Anisotropic Metals," Proceedings of the Royal Society, Series A, Vol. 193, 1948.

29. Puppo, A. H. and Evensen, H. A., "Strength of Anisotropic Materials Under Combined Stresses," AIAA/ASME 12th Structures, Structural Dynamics and Materials Conference, Anaheim, California, April 19-21.

30. Tsai, Stephen W., "Strength Theories of Filamentary Structures," in R. T. Schwartz and H. S. Schwartz (eds.), Fundamental Aspects of Fiber Reinforced Plastic Composites, Wiley Interscience, New York, 1968, pp. 3-11.

31. Tsai, S. W. and Hahn, H. T., "Failure Analysis of Composite Materials," Inelastic Behavior of Composite Materials, (C. T. Herakovich, ed.), ASME, New York, 1975.

32. Petit, P. H., and Waddoups, E. M., "A Method of Predicting the Non-linear Behavior of Laminated Composites," Journal of Composite Materials, January 1969.

33. Sandhu, R. S., "Ultimate Strength Analysis of Symmetric Laminates," Technical Report AFFDL-TR-73-137, Air Force Flight Dynamics Laboratory, Wright-Patterson Air Force Base, Ohio.

34. Yeow, Y. T. and Brinson, H. F., "An Experimental Investigation on the Tensile Moduli and Strengths of Graphite/Epoxy Laminates," Experimental Mechanics, Nov. 1977, pp. 401-408.

35. Dillard, D. A., "Creep and Creep Rupture of Laminated Graphite/Epoxy Composites," Ph.D. Thesis, March 1981. Also, VPI-E-81-3.

36. Waddoups, M. E., Eisenmann, J. R., and Kaminski, B. E., "Macroscopic Fracture Mechanics of Advanced Composite Materials," J. Comp. Mat., Vol. 5, 1971, pp. 446-454.

37. McClintock, F. A. and Irwin, G. R., "Plasticity Aspects of Fracture Mechanics," ASTM-STP 381, American Society for Testing and Materials, 1965, pp. 84-113.

38. Nuismer, R. J. and Whitney, J. M., "Uniaxial Failure of Composite Laminates Containing Stress Concentrations," ASTM-STP 593, American Society for Testing and Materials, 1975, pp. 117-142.

39. Zweben, C., "Fracture Mechanics and Composite Materials: A Critical Analysis," ASTM-STP 521, American Society for Testing and Materials, 1973, pp. 65-97.

40. Rosen, B. W., Kulkarni, S. V. and McLaughlin, P. V., Jr.,
     "Failure and Fatigue Mechanisms in Composite Materials," in
     Inelastic Behavior of Composite Materials, AMD - Vol. 13, ASME,
     1975, pp. 17-72.

41. Kanninen, M. F., Rybicki, and Griffith, W. I., "Preliminary
     Development of a Fundamental Analysis Model for Crack Growth in a
     Fiber Reinforced Composite Material," ASTM-STP 617, American
     Society for Testing and Materials, 1977, pp. 53-69.

42. Cruse, T. A., "Tensile Strength of Notched Composites," J. Comp.
     Mat., Vol. 7, 1973, pp. 218-229.

43. Snyder, M. D. and Cruse, T. A., "Boundary-Integral Equation
     Analysis of Cracked Anisotropic Plates," Int. J. Fracture, Vol.
     11, 1975, pp. 315-328.

44. Yeow, Y. T., Morris, D. H., and Brinson, H. F., "The Fracture
     Behavior of Graphite/Epoxy Laminates," Experimental Mechanics,
     Vol. 19, No. 1, Jan. 1979, pp. 1-8.

45. Yeow, Y. T., Morris, D. H., and Brinson, H. F., "A Correlative
     Study Between Analysis and Experiment on the Fracture Behavior of
     Graphite/Epoxy Laminates," J. of Testing & Eval., Vol. 7, No. 2,
     1979.

46. Kanninen, M. F., Rybicki, and Brinson, H. F., "A Critical Look at
     Current Applications of Fracture Mechanics to the Failure of
     Fiber-Reinforced Composites," Composites, Jan. 1977, pp. 17-27.

47. Morris, D. H. and Hahn, H. T., "Mixed-Mode Fracture of Graphite/
     Epoxy Composites: Fracture Strength," J. Comp. Mat., Vol. 11,
     1977, pp. 124-138.

48. Dally, J. W. and Alfirevich, I., "Application of Birefringent
     Coatings to Glass-Fiber-Reinforced Plastics," Experimental
     Mechanics, Vol. 9, March 1969.

49. Schapery, R. A., "Stress Analysis of Viscoelastic Composite
     Materials," in Composite Materials Workshop, S. W. Tsai, J. C.
     Halpin and N. J. Pagano, eds., Technomic Publishing Co., 1968.

50. Morris, D. H., Brinson, H. F., and Yeow, Y. T., "The Viscoelastic
     Behavior of the Principal Compliance Matrix of a Unidirectional
     Graphite/Epoxy Composite," Polymer Composites, Sept. 1980, Vol. 1,
     No. 1, pp. 32-36. Also, VPI-E-79-9, Feb. 1979.

51. Yeow, Y. T., Morris, D. H., and Brinson, H. F., "The Time-
     Temperature Behavior of a Unidirectional Graphite/Epoxy Laminate,"
     Composite Materials: Testing and Design (5th Conference), STP
     674, ASTM, Phil., 1979, pp. 263-281. Also, VPI-E-78-4, Feb. 1978.

52. Brinson, H. F., Griffith, W. I., and Morris, D. H., "Creep Rupture of Polymer-Matrix Composites," Proceedings, Fourth International Congress on Experimental Stress Analysis, and Experimental Mechanics, in press. Also, VPI-E-80-18, July 1980.

53. Ferry, J. D., Viscoelastic Properties of Polymers, John Wiley & Sons, NY, 1970.

54. Lohr, J. J., "Yield Master Curves for Various Polymers Below Their Glass Transition Temperature," Transactions of the Society of Rheology, Vol. 9, No. 1, 1965.

55. Markovitz, H., "Superposition in Rheology," J. Polymer Science, Symposium No. 50, 1975.

56. Christensen, R. M., Theory of Viscoelasticity, Academic Press, N.Y., 1971.

57. McCrum, N. G. and Pogany, G. A., "Time-Temperature Superposition in the Alpha Region of an Epoxy Resin," J. Macromolecular Science - Phys., B4(1), 1970.

58. Daugste, C. L., "Joint Application of Time-Temperature and Time-Stress Analogies to Constructing Unified Curves," Polymer Mechanics, Vol. 10, No. 3, 1974, pp. 359-362.

59. Crossman, F. W. and Flaggs, D. L., LMSC-D33086, Lockheed Palo Alto Research Laboratory, November 1978.

60. Griffith, W. I., Morris, D. H., and Brinson, H. F., "The Accelerated Characterization of Viscoelastic Composite Materials," VPI-E-80-15, April 1980.

61. Griffith, W. I., Morris, D. H., and Brinson, H. F., "Accelerated Characterization of Graphite/Epoxy Composites," Proceedings of the Third International Conference on Composite Materials, Palais des Congrès, Paris, France, Aug. 25-30, 1980, in press. Also, VPI-E-80-27, Sept. 1980.

62. Schapery, R. A., "On the Characterization of Non-Linear Viscoelastic Materials," Polymer Engineering and Science, Vol. 9, No. 4, 1969.

63. Lou, Y. C. and R. A. Schapery, "Viscoelastic Characterization of a Nonlinear Fiber-Reinforced Plastic," Journal of Composite Materials, Vol. 5, 1971.

64. Beckwith, S. W., "Viscoelastic Characterization of a Nonlinear,
     Glass-Epoxy Composite Using Micromechanics Theory," presented at
     Annual Meeting of Jannaf, San Francisco, Feb. 1975.

65. Cartner, J. S., W. I. Griffith, and H. F. Brinson, "The Visco-
     elastic Behavior of Composite Materials for Automotive Applica-
     tions," Composite Materials in the Automobile Industry, ASME, 1978.

66. Cartner, J. S. and H. F. Brinson, "The Non-Linear Viscoelastic
     Behavior of Adhesives and Chopped Fiber Composites," VPI-E-78-21,
     1978. (Also, VPI&SU M.S. Thesis of J. S. Cartner.)

67. Crochet, M. J., "Symmetric Deformations of Viscoelastic-Plastic
     Cylinders," Journal of Applied Mechanics, Vol. 33, 1966, pp. 327-
     334.

68. Nagdi, P. M. and Murch, S. A., "On the Mechanical Behavior of
     Viscoelastic-Plastic Solids," Journal of Applied Mechanics, Vol.
     30, 1963, p. 321.

69. Brinson, H. F., "The Viscoelastic-Plastic Characterization of a
     Ductile Polymer," Deformation and Fracture of High Polymers,
     H. Kausch, et al., eds., Plenum Press, New York, 1974.

70. Heimbach, R. A. and Sanders, B. A., "Mechanical properties of
     automotive chopped fiber reinforced plastics," Composite Materials
     in the Automobile Industry, ASME, 1978.

71. Conway, J. B., Stress-Rupture Parameters: Origin Calculation and
     Use, Gordon and Breach, N.Y., 1969.

72. Landel, R. F. and Fedors, R. F., "Rupture of Amorphous Unfilled
     Polymers," Fracture Processes in Polymeric Solids, B. Rosen, ed.
     Interscience Publishers, N.Y., 1964.

73. Rowlands, R. E., "Flow and Failure of Biaxially Loaded Composites:
     Experimental Theoretical Correlation," Inelastic Behavior of
     Composite Materials, ASME, 1975.

74. Brüller, O. S., "On the Damage Energy of Polymers in Creep,"
     Polymer Engineering and Science, Vol. 18, No. 1, 1979.

75. Morris, D. H., Brinson, H. F., Griffith, W. I., and Yeow, Y. T.,
     "The Viscoelastic Behavior of a Composite in a Thermal Environment,"
     in Severe Environments (D. P. H. Hasselman and R. A. Heller, eds.),
     Plenum Press, NY, 1980, pp. 693-707. Also, VPI-E-79-40, Sept. 1979.

76. Dillard, D. A., Morris, D. H., and Brinson, H. F., "Predicting Viscoelastic Response and Delayed Failures in General Laminated Composites," 6th Conference on Composite Materials Testing and Design, Phoenix, AZ, May 12-13, 1981.

TEACHING AND RESEARCH IN EXPERIMENTAL MECHANICS.

COMPLEMENTARITY OF THEORY AND EXPERIMENT

Vagn Askegaard
Technical University of Denmark
Structural Research Laboratory
2800 Lyngby

1. Introduction

This part of the book describes an attempt made at the Structural
Research Laboratory of the Technical University of Denmark to formulate
a course that emphasizes the fundamental relationship between theory and
experiment and thus between teaching and research in the thermo-mechanical
field.

The text is part of the book "Eksperimentel Mekanik", by Vagn Askegaard and Jørgen Nielsen, which has been published by Polyteknisk Forlag,
Copenhagen, and which is used as a textbook in the last part of a 5-year
curriculum leading to a master's degree in structural engineering.

The relationship between theory and experiment has been discussed
by several authors, for example, Leipholz.[1], Drucker.[2] and Pindera.[3]. It

can be summarized as follows.

A theory is a simplified description of a phenomenon - simplified because the effect of a large part of the events in Nature on the phenomenon is unknown and because the theory would become completely unwieldy and therefore be of considerably less value if account had to be taken of all such events. The theory and the experiments serve to pinpoint the most important of these events.

Theory and experiment alternate in importance during the development of a discipline. Initially, experiments are performed to collect data, which form the basis for the formulation of assumptions on which the theory is then constructed in such a way that it can be used to predict what is going to happen in an imaginary situation. Experiments are then used once more in order to try to demonstrate that the theory formulated describes the phenomenon within the entire area covered by the assumptions.

If agreement is found between the results of the tests and the theoretical values, the theory is regarded as applicable within the specified realm of validity. If the theory is simple to use, tests in this area will then be superfluous. If, however, the theory is difficult to use, special cases will have to be solved by tests, and these will, of course, also be necessary if an attempt is made to apply the theory beyond the limits determined by its assumptions.

In the text, the relationship between theory and experiment is illustrated by examples which show that the use of theories is necessary in connection with the planning of tests and assessment of measuring errors. In addition, some of the assumptions frequently used in the formulation of theories and the experimental possibilities of verifying such assumptions are discussed.

Chapter 2 gives a short account of general and specialized equation systems used for describing thermo-mechanical phenomena.

The general equations for a 3-dimensional body are taken as the starting-point because they describe the broadest class of problems and can therefore always be used if a more specialized equation system does not suffice.

In chapter 3, these equation systems are used for the formulation of model laws, which are necessary in the planning of model tests and for an evaluation of the results of these.

In chapter 4, an attempt is made to show how theoretical considerations can - and must - be used to determine how the measuring operation itself affects the quantity to be measured.

Finally, chapter 5 contains a discussion of the validity of some of the assumptions on which the theories are constructed and of how appropriate experiments can be used to produce the information required to establish such assumptions.

2. The Basic Equations of Continuum Mechanics and Equation Systems of Derived Theories

2.1 Basic equations   The foundation of continuum mechanics rests on the laws of physics which describe observable events with a greater or lesser degree of accuracy.

The number of phenomena that influences the event to be described
determines how many physical laws have to be included in the elaboration
of the continuum theory. The more laws that have to be included, the more
complicated the theory becomes.

A continuum theory that describes thermo-mechanical phenomena will
cover a large part of the events within civil and mechanical engineering.
Such a general theory is based on the following hypotheses (physical laws):

A: Conservation of mass.

B: The rate of change of momentum is equal to the applied
    force. (Newton's laws).

C: The rate of change of angular momentum is equal to the
    total applied torque.

D: Balance of energy (1st law of thermodynamics).

E: Principle of entropy (2nd law of thermodynamics).

Hypothesis E implies that the time rate of change of the total entropy
is never less than the sum of the influx of entropy through the surface
and the entropy supplied by the body sources.

The above-mentioned hypotheses lead to a system of field equations
which, together with the boundary conditions (geometry, load distribution,
volume forces, heat influx etc.) and initial conditions, describes the
event. The following field equations are obtained, cf. for instance,
Eringen.[4] .

A) The conservation of mass hypothesis leads to the continuity
equation:

$$\frac{\partial \rho}{\partial t} + (\rho v_i)_{,i} = 0 \tag{2.1}$$

or $\quad \overset{\centerdot}{\rho} + \rho v_{i,i} = 0$

Here ρ is the specific weight and v_i are the components of the velocity vector at the point considered, in relation to a coordinate system fixed in space. $\frac{\partial f}{\partial t}$ is the time derivative at a fixed point in space, while $\overset{\centerdot}{f}$ is the time derivative at a point in the body.

B) and C) Hypotheses B and C leads to the equations of motion. For a non-polar medium, i.e., a medium with a symmetrical stress tensor, the equations are:

$$\sigma_{ij,j} + \rho(f_i - \overset{\centerdot}{v}_i) = 0$$

or $\tag{2.2}$

$$\sigma_{ij,j} + \rho(f_i - \frac{\partial v_i}{\partial t} - v_j v_{i,j}) = 0$$

$$\sigma_{ij} = \sigma_{ji} \tag{2.3}$$

f_i being the acceleration in the field of gravity.

D)   The first law of thermodynamics states that the time rate of change of the internal energy U and the kinetic energy is equal to the sum of the rate of work done by external forces and the energy rate of other sources. Considering only thermomechanical phenomena, the only other sources are heat sources.

This theorem can be expressed as follows (energy equation):

$$\rho \overset{\centerdot}{U} = \frac{1}{2} \sigma_{ij}(v_{i,j} + v_{j,i}) + \rho h - q_{k,k} \tag{2.4}$$

U is the internal energy per unit mass (elastic or non-elastic energy and of stored heat energy, accessible or not). h is the rate of heat energy produced per unit mass and q_i is the heat flow vector representing the flow of heat per unit area out of a surface.

The system of equations (2.1) - (2.4) is valid for any continuum, e.g. gases, liquids or solids. To solve the equations, information about boundary - and initial conditions is required. There are 17 unknowns in the 8 equations (2.1) - (2.4), viz., ρ , v_i , σ_{ij} , q_i and U . f_i and h are supposed to be prescribed as part of the boundary conditions. Nine additional equations must then be formulated to equal the number of equations and unknowns required for solving a problem.

The second law of thermodynamics, where entropy and absolute temperature are introduced, adds an inequality to the equation system. This inequality is important for establishing a constitutive theory where restrictions on the form of a set of equations termed constitutive equations are derived. These equations describe the material behaviour and represent the additional equations. The inequality may take the following form, cf. for instance Eringen.[4] .

E) $\rho(\dot{\eta} - \frac{\dot{U}}{\theta}) + \frac{1}{\theta}\,\sigma_{ij}v_{j,i} + \frac{1}{\theta}\,q_k(\ln\theta)_{,k} \geq 0$ (2.5)

where η is the entropy density and θ is the absolute temperature.

The axioms introduced in the development of a constitutive theory include the following:

The quantities σ_{ij} , q_i , U and η at a point in the body A at time t are uniquely related to the history of all points in A , the history being characterized by the point's motion x_i' and its temperature θ' in the time interval $-\infty < t' \leq t$.

This axiom, called the axiom of determinism, leads to the following very general formulation of the constitutive equations:

$$
\begin{aligned}
\sigma_{ij} &= \sigma_{ij}(x_k',\theta',t') \\
q_i &= q_i(x_k',\theta',t') \qquad x_k' \in A,\ x_k \in A \\
U &= U(x_k',\theta',t') \qquad \text{and } -\infty < t' \leq t \\
\eta &= \eta(x_k',\theta',t')
\end{aligned}
\qquad (2.6)
$$

where the motion x_k corresponds to the time t , x_k and t being the independent variables.

Two new unknown quantities, viz., the entropy density and the absolute temperature, are introduced in (2.6), bringing the number of unknowns up to 19. By adding the eleven constitutive equations given by (2.6) to the equations (2.1) - (2.4) the number of equations is also seen to be 19.

Simplification of equations (2.6) is aimed at by adding further axioms. Usually the axiom of neighborhood is introduced, saying that the dependent variables σ_{ij} , U , q_i and η are only functions of x_k and θ in the immediate neighborhood of the observed point, such that σ_{ij} , q_i , U and η may be expressed as functions of motion and temperature and their gradients of first and higher order at the point of observation. If only gradients of the first order are included in the description the material is called simple. Another axiom is the axiom of objectivity, which states that the constitutive equations must be form-invariant with respect to rigid motions of the spatial frame of reference. By using this axiom x_k'

is eliminated from the constitutive equations such that:

$$\sigma_{ij} = \sigma_{ij}(x'_{k,l}, x'_{k,lm} \cdots \theta', \theta'_{,k}, \theta'_{,kl} \cdots t') \text{ etc.}$$

As $x'_{k,l}$ corresponds to the strain tensor e_{kl} we have:

$$\sigma_{ij} = \sigma_{ij}(e'_{kl}, e'_{kl,m} \cdots \theta', \theta'_{,k}, \theta'_{,kl} \cdots t') \text{ etc.}$$

For a simple material we get:

$$\sigma_{ij} = \sigma_{ij}(e'_{kl}, \theta', \theta'_{,k}, t') \text{ etc.}$$

The effect of introducing these and other axioms on the form of equations (2.6) is discussed in the literature.

By adding a sufficient number of assumptions, the equations (2.6) change from functionals to closed expressions, which are necessary for an analytical or numerical solution of equations (2.1) - (2.6). This simplification of equations (2.6) is followed by a reduction in the field of application.

In the following, examples are given of theories resulting from introducing assumptions with restricted validity.

2.2 Fluids   The behaviour of a large number of compressible fluids is often considered to be well described by the following extra set of assumptions:

1)   A fluid has no memory, which means that no account
     need to be taken of particle history.

2)   A linear relationship exists between components in the s_{ij}
     and $(v_{i,j} + v_{j,i})$ tensors where $s_{ij} = \sigma_{ij} + p\delta_{ij}$ and where

the scalar p is called the pressure. This relationship takes the form $s_{ij} = K_{ijpq}(v_{p,q} + v_{q,p})$ where K_{ijpq} are viscosity coefficients.

3) The material is isotropic. The relationship mentioned in 2) now becomes $s_{ij} = \gamma\delta_{ij}v_{k,k} + \mu(v_{i,j} + v_{j,i})$ where γ and μ are viscosity coefficients.

4) A kinetic equation of state exists of the form $p = p(\rho,\theta)$, where θ is the absolute temperature.

5) Fourier's law of heat conduction in an isotropic material $q_i = -\lambda\theta,i$ is valid. λ is called the coefficient of thermal conductivity.

6) A caloric equation of state exists of the form $U = U(\rho,\theta)$

Using this set of extra assumptions, we obtain the equation system (2.7). The equation system should be supplemented with the boundary - and initial conditions characterizing the actual problem.

$$\left.\begin{array}{l}
\dot{\rho} + \rho v_{i,i} = 0 \\[4pt]
\sigma_{ij,j} + \rho(f_i - \dot{v}_i) = 0 \\[4pt]
\rho\dot{U} = \frac{1}{2}\sigma_{ij}(v_{i,j} + v_{j,i}) + \rho h - q_{k,k} \\[4pt]
\sigma_{ij} = -p\delta_{ij} + \gamma\delta_{ij}v_{k,k} + \mu(v_{i,j} + v_{j,i}) \\[4pt]
p = p(\rho,\theta) \\[4pt]
q_i = -\lambda\theta,_i \\[4pt]
U = U(\rho,\theta)
\end{array}\right\} \qquad (2.7)$$

A total of 16 unknowns ρ , v_i, σ_{ij}, q_i, U, θ and p corresponds to the number of equations. The entropy density η could be added to the number of unknowns by adding a constitutive equation $\eta = \eta(e_{ij},\theta,\theta,_j,t)$

to the number of equations. In most cases, however, knowledge of η is

not considered to be important and the extra equation is therefore omitted.

The entropy inequality (2.5) must, however, be imagined to be used in the

formulation of the extra set of assumptions.

The fluid may be a gas that can be considered ideal if the tempera-

ture and the pressure do not reach extreme values. The kinetic equation

of state will, in this ideal case, be expressed by $p = \rho R\theta$ where R is

the gas constant, which is a material parameter. The caloric equation of

state will be $U = U(\theta) = \int c_v d\theta + \text{const.}$ where c_v is the specific heat

at constant volume.

If further assumptions are introduced, viz.,

7) Incompressibility and homogeneity i.e. $\rho = \rho_o = \text{const.}$

8) Constant temperature, i.e. $\theta = \text{const.}$ and $q_i = 0$.

then (2.7) can be simplified as shown in (2.8).

$$
\left.
\begin{aligned}
&v_{i,i} = 0 \\[2pt]
&\sigma_{ij,j} + \rho_o(f_i - \dot{v}_i) = 0 \\[2pt]
&\rho_o\dot{U} = \tfrac{1}{2}\,\sigma_{ij}(v_{i,j} + v_{j,i}) + \rho h \\[2pt]
&\sigma_{ij} = -p\delta_{ij} + \mu(v_{i,j} + v_{j,i})
\end{aligned}
\right\} \qquad (2.8)
$$

$$U = U(\theta) = \text{const.}$$

The relation $p = p(\rho,\theta)$ becomes undetermined because of the in-

compressibility which leaves an unknown hydrostatic stress. p can instead

be found by combining the continuity equation and the constitutive equa-

tions expressing σ_{ij}. It is found that $p = -\tfrac{1}{3}\sigma_{ii}$.

Assumption 7 seems to approximate the behaviour of gases and liquids well when the velocity is much below the speed of sound. An exception is waterwaves breaking against a wall where a small amount of dispersed air bubbles influences the pressure distribution.

A more extensive treatment of fluid mechanics is given for instance in Eringen.[4], Liepmann, Roshko.[5], and Landau, Lifshitz.[6].

2.3 Linear elasticity   A simple material is said to be elastic if σ_{ij}, q_i, U and η at a point at time t are functions only of the strain tensor e_{ij} and the temperature θ at that point at time t . There is thus no dependence on the history and the strain at other points of the material.

From axioms in the constitutive theory it follows that q_i must be zero. The equation system for a body of homogenous linear elastic material can be derived from the following extra assumptions se for instance Hodge.[7].

a) the temperature in the body is constant and

   h = 0.

b   the strains are so small that the expressions

$e_{ij} = \frac{1}{2}(u_{i,j} + u_{j,i})$ and $\dot{e}_{ij} = \frac{1}{2}(v_{i,j} + v_{j,i})$ are valid.

c) the internal energy density is uniquely related to the strain tensor in the following way

$$\rho U = a + b_{ij}e_{ij} + \frac{1}{2} c_{ijkl}e_{ij}e_{kl}$$

where a, b and c are material constants.

Assumption c) represents a constitutive condition. If the initial state is chosen so that stress, strain and internal energy are zero, then a = 0 . From $\sigma_{ij} = \frac{\partial U}{\partial e_{ij}}$, which follows from the energy equation, it can be seen that $b_{ij} = 0$. The equation system (2.1) - (2.4) now takes the form

$$\dot{\zeta} + \rho \dot{e}_{ii} = 0 \quad \Rightarrow \quad \rho = \rho_o(1-e_{ii})$$

$$\sigma_{ij,j} + \rho_o(f_i - \ddot{u}_i) = 0$$

$$\sigma_{ij} = \frac{1}{2} c_{ijkl}(u_{k,l} + u_{l,k}) = c_{ijkl} \cdot e_{kl}$$

$$e_{ij} = \frac{1}{2}(u_{i,j} + u_{j,i}).$$

(2.9)

From (2.9) it can be seen that ρ is explicitly determined from the strains, and the strains from the displacements. The problem can thus be reduced to finding the 9 unknowns σ_{ij} and u_i from the equations of motion and the stress-displacement relations in (2.9).

If information about the internal energy is wanted, the quantities found from solving (2.9) can be inserted in the expression in assumption c).

The change of entropy can be shown to be zero.

By introducing the assumption of isotropy the number of independent coefficients c_{ijkl} is reduced to two, for instance, the modulus of elasticity E and Poisson's ratio ν or the Lamé constants λ_L and μ_L. The equation system then takes the form (2.10)

$$\rho = \rho_o(1-e_{ii})$$

$$\sigma_{ij,j} + \rho_o(f_i - \ddot{u}_i) = 0$$

$$\sigma_{ij} = \lambda_L \delta_{ij} e_{kk} + 2\mu_L e_{ij} \quad \text{(Hooke's law)}$$

$$e_{ij} = \frac{1}{2}(u_{i,j} + u_{j,i})$$

(2.10)

The Lamé constants are related to E and ν in the following way:

$$\lambda_L = \frac{E\nu}{(1+\nu)(1-2\nu)} \qquad \mu_L = \frac{E}{2(1+\nu)}$$

(2.11)

2.4 Linear visco-elasticity   The linear theory of visco-elasticity is used to describe the behaviour of a number of liquids and solids
with time-dependent material properties. The constitutive equations expressing the relations between stresses and strains are either given in
integral or differential form such that an increase of the stress by a
factor n in a time period increases the strain by the same factor. Readers interested in the theory of linear visco-elasticity are referred to,
for instance, Gurtin and Sternberg.[8], Fung.[9] and Ferry.[10].

The following set of extra assumptions is introduced:

a) the material is homogeneous and isotropic.

b) the temperature in the body is constant, i.e. $q_i = 0$
    and $h = 0$.

c) the strains are so small that the expressions
    $e_{ij} = \frac{1}{2}(u_{i,j}+u_{j,i})$ and $\dot{e}_{ij} = \frac{1}{2}(v_{i,j}+v_{j,i})$ are valid.

d) the body is not loaded until time $t = 0$.

e) Boltzmann's superposition law is valid.

In the uniaxial stress state it takes the form

$$e(t) = \sum_{0}^{n} \Delta\sigma_i k(t-\tau_i) \rightarrow \int_{0}^{t} k(t-\tau) \frac{\partial\sigma}{\partial\tau} d\tau$$

where $k(t)$ is a time-dependent material function.

A generalization of this expression leads to:

$$e_{ij} = \int_{0}^{t} k_{ijmn}(t-\tau) \frac{\partial\sigma_{mn}}{\partial\tau} d\tau.$$

The number of time-dependent material functions $k_{ijmn}(t)$ reduces
to two when the material is isotropic.

The extra assumptions a - e lead to the following equation system

for linear visco-elastic phenomena:

$$\dot{\rho} + \rho v_{i,i} = 0 \;\rightarrow\; \rho = \rho_0(1-e_{ii})$$

$$\sigma_{ij,j} + \rho_0(f_i - \ddot{u}_i) = 0$$

$$e_{ij} = \tfrac{1}{2}(u_{i,j} + u_{j,i})$$

I $$\sigma_{ij} = \int_0^t [G_1(t-\tau)(\frac{\partial e_{ij}(\tau)}{\partial \tau} - \tfrac{1}{3}\delta_{ij}\frac{\partial e_{kk}(\tau)}{\partial \tau}) + \tfrac{1}{3}\delta_{ij}G_2(t-\tau)\frac{\partial e_{kk}(\tau)}{\partial \tau}]d\tau$$

II $$e_{ij} = \int_0^t [J_1(t-\tau)(\frac{\partial \sigma_{ij}(\tau)}{\partial \tau} - \tfrac{1}{3}\delta_{ij}\frac{\partial \sigma_{kk}(\tau)}{\partial \tau} + \tfrac{1}{3}\delta_{ij}J_2(t-\tau)\frac{\partial \sigma_{kk}(\tau)}{\partial \tau}]d\tau$$

III $$a_0 + a_1\frac{\partial \sigma'_{ij}}{\partial t} + \dots a_{n_1}\frac{\partial^{n_1}\sigma'_{ij}}{\partial t^{n_1}} = b_0 + b_1\frac{\partial e'_{ij}}{\partial t} + \dots b_{m_1}\frac{\partial^{m_1}e'_{ij}}{\partial t^{m_1}},$$

$$c_0 + c_1\frac{\partial \sigma_{kk}}{\partial t} + \dots c_{n_2}\frac{\partial^{n_2}\sigma_{kk}}{dt^{n_2}} = d_0 + d_1\frac{\partial e_{kk}}{\partial t} + \dots d_{m_2}\frac{\partial^{m_2}e_{kk}}{\partial t^{m_2}}$$

(2.12)

Information about boundary and initial conditions for the actual

problem should be added to the equation system.

This is seen to consist of 16 equations and 16 unknowns. The consti-

tutive equations are given in three alternative forms I, II and III,

where the most suitable for solving the actual problem should be

chosen. $G_1(t)$ and $G_2(t)$ are called relaxation functions, $J_i(t)$

and $J_2(t)$ are creep functions and $E(t)$ is a time dependent modulus

of elasticity and $\nu(t)$ a time dependent Poisson's ratio.

From (2.12)III it can be seen that if a homogeneous state of stress

is produced in a test specimen at time $t = 0$ and thereafter kept constant,

the following homogeneous state of strain is obtained:

$$e_{ij} = -\frac{\nu(t)}{E(t)}\delta_{ij}\sigma_{kk} + \frac{1+\nu(t)}{E(t)}\sigma_{ij}$$

(2.13)

It is thus possible to determine E(t) and ν(t) from a uniaxial
test.

   2.5 Calculation - Experiment   The same phenomenon can often be
described by several theories that are only separated by number of assumptions made, and it can thus be said that one theory simply slides into
another,

   It would hardly be warranted to advocate any particular theory at the
expense of the others, since the general principle must be to choose the
simplest possible theory that describes the phenomenon in question reasonably well. The choice must be based on a critical study of the assumptions
of the theories. In case of doubt, a theory based on fundamental physical
observations should be preferred. Choosing a theory, whose assumptions are
judged to be satisfactory does not necessarily mean that a practical problem can be solved theoretically. Calculating problems in connection with
analytical solutions often make it necessary for us to limit ourselves to
the solution of problems in which highly simplified assumptions are permissible, e.g.

   Isotropy in material properties.

   Linear elasticity or linear visco-elasticity.

   Simple geometry.

   Simple loading cases.

   These assumptions are satisfactory for the solution of a number of
technically important problems in connection with beams, columns, frames,
plates and shells of steel, concrete or wood. Numerical methods, e.g.
finite element method, enable the introduction of more complicated mate-

rial descriptions, including failure criteria and also make it possible
to solve problems involving complicated geometry. With this method the
practical solution problems increase with the number of elements (difficult geometry), the number of necessary iterations (time-dependent phenomena) and if there are non-linearities in the set of equations. Readers
interested in numerical methods are referred to Zienkiwicz, Cheung.[11] and
Stanley.[12].

The more uncertain the foundation on which the analysis is based,
the more reason there is to observe the completed structures, and to acquire supplementary information by performing other experiments.

Even though observations on the structure itself play a much larger
role than observations on models, model techniques still have an important
role to play.

The transfer of results from model to structure requires knowledge
of the system of equations that governs the phenomenon, or knowledge of
all the parameters contained in the equation system (dimensional analysis).

Model theory is treated in the following chapter, from which it will
also be seen that conflicting requirements to scale ratios and the practical possibilities of ensuring the desired material properties and loads
limit the use of model tests. For example, model tests cannot be used
when an unknown initial field governs the behaviour of the structure
(shrinkage-rolling-welding stresses), or when the structural material has
time-dependent properties and the structure is dynamically loaded.

In these and other important cases, direct measurements must be
performed on the actual structure.

3. Model Laws from Systems of Equations

Model tests are characterized by the field (stress, strain, displacement, temperature etc.) in a structure, A, being determined by measuring
the corresponding field in a model, B .

The following procedure can be used for systematic formulation of
model laws:

1) An equation system describing the field in A and B is established on the basis of reasonable assumptions.

2) The problem now is to determine the conditions that must be established to ensure that the system of equations for B is satisfied
   by a field which can be simply expressed by means of the field in
   A .

   The relationship between the two fields is to be characterized by constant ratios (scale ratios) between corresponding quantities in A and B where the contemplated quantities characterize
   the field. The scale ratios are characterized by the letter K
   with index for the quantity contemplated.

   $K_\sigma = \dfrac{\sigma_{ij}^A}{\sigma_{ij}^B}$ thus characterizes the scale ratio for the components

   in the stress tensor σ_{ij} . The length scale ratio is characterized by K_1 .

3) If the established conditions can be satisfied, it will be possible to determine a quantity in A by measuring the corresponding
   quantity in B multiplying this quantity by the scale ratio.

4) If conflicting requirements arise, then it must be decided whether

it is permissible to alter the assumptions to get rid of the con-
flicts. This introduces a scale error, and the magnitude of this
must be estimated.

The above-mentioned procedure is illustrated by the following example.

3.1   Structure made of a material with unknown properties   If the
structure is assumed to be made of a material with unknown properties, the
field will have to be described by the basic equations.

If the structure is designated A , then the equation system is as
follows, see section 2.1:

$$
\left.\begin{array}{l}
\dot{\rho}^A + \rho^A\, v^A_{i,i} = 0 \\[2mm]
\sigma^A_{ij,j} + \rho^A(f^A_i - \dot{v}^A_i) = 0 \\[2mm]
\rho^A\, \dot{U}^A = \frac{1}{2}\, \sigma^A_{ij}(v^A_{i,j}+v^A_{j,i}) + \rho^A\, h^A - q^A_{k,k} \\[2mm]
\sigma^A_{ij} = \sigma^A_{ij}(x^{A'}_k,\theta^{A'},t^{A'}) \qquad\qquad U^A = U^A(x^{A'}_k,\theta^{A'},t^{A'}) \\[2mm]
q^A_i = q^A_i(x^{A'}_k,\theta^{A'},t^{A'}) \qquad\qquad \eta^A = \eta^A(x^{A'}_k,\theta^{A'},t^{A'}) \\[2mm]
x^{A'}_k \in A, \quad x^A_k \in A \ \ \text{and} \ \ -\infty < t^{A'} \leq t^A
\end{array}\right\}
\qquad (3.1)
$$

If the structure consists of several sub-elements, A_i , then the
above-mentioned system of equations will have to be established for each
of these. The boundary conditions, i.e. stress, displacement, temperature,
heat energy per time unit, etc. on the outer edge of the structure will
usually be known, together with the volume force distribution and the distribution of heat sources.

In the above-mentioned formulation of the constitutive equations, the
magnitudes of σ, q, U and η at any point depend on the motion and tem-

perature of all the points in the body in the time interval from $-\infty$ to the time of observation.

A geometrically similar model, designated B , is now made from a material that is assumed to be identical to the material in A , i.e. a material with the same, but unknown, constitutive equations F_α . The only thing known is the parameters that are included in the description. The only possible way to model is thus to choose the same material. Corresponding basic equations can be established for the model by changing index A to B in (3.1).

That the materials are identical means that the parameters at a certain point cannot, as previously, be regarded as a function of $x_i' - \theta' - t'$ history throughout the whole body, unless the length scale ratio is one, which is a trivial case.

By introducing the axiom of neighborhood and objectivity the constitutive equations take the following form (see section 2.1)

$$
\left.
\begin{aligned}
\sigma_{ij} &= \sigma_{ij}(e_{kl}',\ e_{kl,m}'\cdots\cdots,\theta',\theta_{,k}',\theta_{,kl}'\cdots t') \\[6pt]
q_i &= q_i\ (e_{kl}',\ e_{kl,m}'\cdots\cdots,\theta',\theta_{,k}',\theta_{,kl}'\cdots t') \\[6pt]
U &= U\ (e_{kl}',\ e_{kl,m}'\cdots\cdots,\theta',\theta_{,k}',\theta_{,kl}'\cdots t') \\[6pt]
\eta &= \eta\ (e_{kl}',\ e_{kl,m}'\cdots\cdots,\theta',\theta_{,k}',\theta_{,kl}'\cdots t') \\[6pt]
x_k &\in A \text{ or } B \qquad -\infty < t' \le t
\end{aligned}
\right\}
\qquad (3.2)
$$

Again these equations will only satisfy the requirements of being identical in A and B if the length scale ratio is one. This is because the equations include gradients of first and higher order in strain and temperature.

It must therefore be assumed that the material properties can be expressed by constitutive equations of the following form:

$$\sigma_{ij} = \sigma_{ij}(e'_{kl}, \theta', t')$$
etc.

However, experiments show that q_i depend on temperature gradients.
This therefore means that model tests with identical materials in A and
B cannot be carried out when heat conduction occurs. It must thus be assumed in the following that the temperature in the body is constant, or
that it varies so slowly that q_i is negligible. This again means that
the heat source energy h and the transformation of mechanical work to
heat must be insignificant. (3.1) thus goes into (3.3)

$$\left. \begin{array}{l} \dot{\rho}^A + \rho^A v^A_{i,i} = 0 \\[2mm] \sigma^A_{ij,j} + \rho^A(f^A_i - \dot{v}^A_i) = 0 \\[2mm] \rho^A \dot{U}^A = \frac{1}{2} \sigma^A_{ij}(v^A_{i,j} + v^A_{j,i}) \\[2mm] \sigma^A_{ij} = \sigma^A_{ij}(e^{A'}_{kl}, \theta^{A'}, t^{A'}) \qquad U^A = U^A(e^{A'}_{kl}, \theta^{A'}, t^{A'}) \\[2mm] q^A_i = q^A_i(e^{A'}_{kl}, \theta^{A'}, t^{A'}) \qquad \eta^A = \eta^A(e^{A'}_{kl}, \theta^{A'}, t^{A'}) \end{array} \right\} \qquad (3.3)$$

$$x^A_k \in A \quad \text{and} \quad -\infty < t'^A \le t^A$$

The question now is whether the similar equation system for B is
satisfied by a field which, by using the scale ratios, can be expressed
by the field that satisfies (3.3).

Introducing the scale ratios $K_\rho = \dfrac{\rho^A}{\rho^B}$, $K_l = \dfrac{l^A}{l^B}$ etc. in the equation
system for B , the following expressions are obtained

$$\dot{\rho}^B + \rho^B v^B_{i,i} = 0 \rightarrow \frac{K_t}{K_\rho} \dot{\rho}^A + \frac{K_1 K_t}{K_\rho K_1 K_e} \rho^A v^A_{i,i} =$$

$$\frac{K_t}{K_\rho} (\dot{\rho}^A + \rho^A v^A_{i,i}) = 0 \quad \underline{\text{if} \quad K_e = 1}$$

$$\sigma^B_{ij,j} + \rho^B(f^B_i - \dot{v}^B_i) = 0 \rightarrow \frac{K_1}{K_\sigma} \sigma^A_{ij,j} + \frac{1}{K_\rho} \rho^A (\frac{1}{K_f} f^A_i - \frac{K^2_t}{K_1 K_e} \dot{v}^A_i) =$$

$$\frac{K_1}{K_\sigma} (\sigma^A_{ij,j} + \rho^A(f^A_i - \dot{v}^A_i)) = 0 \quad \underline{\text{if} \quad \frac{K_1}{K_\sigma} = \frac{1}{K_\rho K_f} = \frac{K^2_t}{K_\rho K_1 K_e}}$$

$$\rho^B \dot{U}^B = \frac{1}{2} \sigma^B_{ij}(v^B_{i,j} + v^B_{j,i}) \rightarrow \frac{K_t}{K_\rho K_U} \rho^A \dot{U}^A =$$

$$\frac{1}{2} \frac{K_1 K_t}{K_\sigma K_1 K_e} \sigma^A_{ij}(v^A_{i,j} + v^A_{j,i}) \rightarrow$$

$$\rho^A \dot{U}^A = \frac{1}{2} \sigma^A_{ij}(v^A_{i,j} + v^A_{j,i}) \quad \underline{\text{if} \quad \frac{K_t}{K_\rho K_U} = \frac{K_1 K_t}{K_\sigma K_1 K_e}}$$

$$\sigma^B_{ij}(e^{B'}_{kl}, \theta^{B'}, t^{B'}) = \sigma^A_{ij}(e^{A'}_{kl}, \theta^{A'}, t^{A'})$$

$$q^B_i(e^{B'}_{kl}, \theta^{B'}, t^{B'}) = q^A_i(e^{A'}_{kl}, \theta^{A'}, t^{A'})$$

$$U^B(e^{B'}_{kl}, \theta^{B'}, t^{B'}) = U^A(e^{A'}_{kl}, \theta^{A'}, t^{A'})$$

$$\eta^B(e^{B'}_{kl}, \theta^{B'}, t^{B'}) = \eta^A(e^{A'}_{kl}, \theta^{A'}, t^{A'})$$

i.e. $K_\sigma = K_q = K_U = K_\eta =$ and $K_\rho = 1$

$$\underline{\text{if } K_t = K_\theta = K_e = 1.}$$

(3.4)

If the underlined conditions are not inconsistent, and if the boundary conditions at B can be adjusted so that they also satisfy the scale ratios in relation to the known boundary conditions at A , then the relationship between the field in A and that in B at an arbitrary set of corresponding points is expressed by the scale ratios.

If all the conditions between the scale ratios in (3.4) are to be satisfied, it is obvious that this will only be possible if the length scale
ratio $K_1 = 1$, i.e. if A and B are identical.

Several more assumptions must thus be introduced if a model test is
to be carried out where $K_1 \neq 1$. It turns out that model tests can be accomplished in the following two cases:

    a) when inertia forces are negligible, i.e. $\overset{A}{\underset{i}{v}} \cong 0$,

      or

    b) when F_α is not dependent on t , i.e. when the material

      has no time-dependent properties.

In a) the set of scale ratios to be used to satisfy the equation
system for B are:

$$K_t = K_\theta = K_e = K_\sigma = K_U = K_\eta = K_\rho = 1 \quad \text{and} \left.\begin{array}{c} \\ \\ \\ \end{array}\right\}$$
$$K_1 = \frac{1}{K_f} \ . \qquad\qquad\qquad\qquad\qquad\qquad\qquad\qquad (3.5)$$

The requirements given in the top line of (3.5) can be satisfied if
$K_e = K_\sigma = K_t = 1$ are fulfilled on the boundary and the same temperature
is established in A and B , as mentioned above. It will then be apparent
from the requirements in the lower line that, for instance, the length
scale ratio K_1 can be freely chosen, whereby K_f are determined. It may
be difficult to maintain the scale ratio K_f for the volume forces (force/
mass unit). For example, if the volume force is caused, by the earths'
gravity field in A , then an artificial K_1 times bigger gravity field
must be generated in B .

In b) we get:

$$
\left.\begin{array}{l}
K_\rho = K_\theta = K_e = K_\sigma = K_U = K_\eta = K_v = 1 \\[2mm]
K_1 = K_t = \dfrac{1}{K_f}
\end{array}\right\} \qquad (3.6)
$$

and

The length scale ratio K_1 , for instance, can be freely chosen, after which the time scale ratio and volume force scale ratio can be determined.

----------o----------

The foregoing pinpoints some of the problems that must be expected in the course of model tests.

We must naturally ensure that the assumptions made are satisfied either by relying on the experience of other researchers or by personally carrying out extensive tests. This raises the question of the best way of performing such tests.

As it is often impossible to follow the history of the material, we must also ensure that the initial field is the same in both A and B . The question now is which methods can be used.

Although inconvenient limits have been set regarding the types of phenomena that can be investigated by model tests, even when the materials in A and B are identical, these tests are still valuable because they can be performed without complete knowledge of the material properties and because they in many cases can be taken all the way to failure without any knowledge of the failure criteria.

4.   Fundamental Measurement Problem

For observations to be regarded as fundamental, in the case of measurements on either prototype or model, they must lead to determination of the quantities included in the basic equations.

Making an observation is the same as one of the human senses being activated by supply of energy.

Only in a very few cases are the human senses sensitive and stable enough to register the desired phenomena, and the signal must therefore be transformed (e.g. strain into an electric resistance change in a strain gauge), with subsequent amplification.

This means that a transducer must be inserted with subsequent interference in the field of the structure and thereby interference in the set of equations governing the phenomenon, since there may be alterations in the geometry or in the quantities in the energy equation. These alterations are usually of little significance for the field of the structure as a whole, but locally, they can mean that the field undergoes an alteration, e.g. the phenomena observed alter during the course of observations.

It is therefore a fundamental measuring problem

to find a measuring principle which enables determination

of the desired quantity with a reasonably acceptable error.

In many cases, the altered set of equations can be solved, whereby the magnitude of the observation interference can be found. In other cases, it cannot be solved, even though it can be formulated. Thus an approximate theoretical solution must be used, with subsequent control tests under known conditions (calibration tests).

Lastly, there may be cases where information about essential parameters in the set of equations is missing. This would warrant obtaining more information by experiment. In this case, both approximate solutions and calibration tests can be considered to determine the experimental error. We thus use approximated theories to substantiate the reliability of test results in cases in which exact theories cannot be established so that these results can be used to formulate more exact theories.

This is a procedure giving acceptable results if the approximate solution and the unknown correct solution are identical at some limit, where the interference with the surroundings is negligible and if we work near this limit, where the error is low.

It is thus unimportant if a correct error of say 2% becomes 3% by using the approximation even though this represents a difference of 50%.

Observation of motion (displacement, geometry) can often be carried out by optical methods so that interference in the field of the structure is negligible.

On the other hand, observation of temperature, strain and stress can often give rise to alteration of the quantity to be observed. No general treatment of such problems can be given, but every problem must be dealt with seperately.

In the following, this problem will be illustrated by detailing the possibilities of measuring stress and strain in a solid body by using inclusions. No account is taken of the errors caused by using a special transducer principle with corresponding measuring equipment. A general treatment of the whole measuring chain has been made by Stein.[13,14] .

4.1 <u>Measurement of stress and strain in solids</u>   It is often desirable to have experimental information concerning the internal stress and
strain field of a body. In this chapter, bodies (matrix) will be discussed
in which cells (inclusions) can be placed, i.e. bodies of granular materials (geotechnical uses) and bodies of materials that can be cast at normal
room temperature (concrete and plastic materials, together with a few low
melting alloys).

The matrix material may be characterized by such simple constitutive
equations that knowledge of one field, e.g. stress field, is sufficient
for determination of the other. If, for instance, the material is linear
elastic, isotropic and homogeneous, a simple relation between stresses and
strains can be obtained. In such cases it will almost always be the strain
field that is measured which is considered to be a relatively simple problem.

If the constitutive equations are more complicated and perhaps even
unknown, then both stresses and strains must be measured to describe the
field.

The field in the inclusion is a function of the field of the body
around the inclusion, and the problem is now to determine the latter by
measuring the former. Measurement of the field in the inclusion is assumed
to be possible.

It can be argued that a stress cell should be flat and very thin and
a strain cell long and very thin to measure a stress component respectively a strain component correctly.

At the limit an approximate solution and an unknown correct solution are identical.

An ellipsoid is now used as the geometrical approximation for the inclusion and linear elasticity and isotropy are taken as approximations for the constitutive equations.

The criteria that have to be fulfilled to allow measurement of a stress component in a stress tensor in a linear elastic matrix material are as follows:

1. The signal measured in the inclusion must be uniquely related to the desired component in the stress tensor, i.e. it must be independent of the other components in the stress tensor.

2. The signal measured must be independent of a stress-free strain field in the inclusion or the matrix (e.g. shrinkage).

3. The signal measured must be independent of the physical constants E and ν in the matrix.

4. Effort must be made to ensure that the presence of the inclusion does change the stress field in the matrix as little as possible when this is loaded or in a stress-free strain field.

The fourth requirement is not necessary when the material is linear elastic, but it is important as stated above because a stress measurement in an arbitrary matrix material cannot be carried out if the requirement is not satisfied in the special case in which the matrix material is linear elastic.

Eshelby.[15] has given expressions for the stress and strain field in an ellipsoidal inclusion.

In Askegaard.[16,17] these expressions are solved numerically for different combinations of E, ν, E_1 and ν_1, where E_1 and ν_1 are the modulus of elasticity and Poissons' ratio in the inclusion, respectively, and E and ν refer to the matrix. Solutions have been found for different ratios between the principal axes in the axisymmetrical ellipsoid. The stress and strain field in an inclusion with a homogenous, stress-free strain field, could it move freely, is also given and is solved for the corresponding combinations. The method is given in the following:

Eshelbys' Expressions   In an infinite medium called a matrix (the Lamé constants λ_L and μ_L) a limited part called inclusion undergoes an alteration of the strain field. This alteration of strain is prescribed such that it would be homogenous if the inclusion could deform freely. However, movement is prevented by the surrounding material. The relationship between the strain and stress field in the matrix and in inclusion is required.

The following procedure is used:

1. The inclusion is imagined to be cut out of the matrix and subjected to the desired homogenous strain field e_{ij}^T. There is thus no stress in the inclusion or the matrix.

2. The inclusion is subjected to the boundary stress distribution $-\sigma_{ij}^T n_j$ where

    $$\sigma_{ij}^T = \lambda_L \, e^T \, \delta_{ij} + 2\mu_L \, e_{ij}^T \quad \text{(Hookes' law)}$$

    ($e^T = e_{ii}^T$. n_j are the components of the normal to the surface).

Thus, the inclusion resumes its original shape and can again be placed in the matrix and fixed to this over the entire surface. The stresses and the strain in the matrix are 0 . The strain in the inclusion is also zero, while the stress field is $-\sigma_{ij}^{T}$.

3. The stress distribution $\sigma_{ij}^{T} n_{j}$ is applied at the imagined interface between the inclusion and the matrix. The desired field is thus produced and is characterized by (4.1).

The strain field in the inclusion and the matrix is denoted e_{ij}^{C} .

The stress field in the matrix is $\sigma_{ij}^{C} = \lambda_{L} e^{C} \delta_{ij} + 2\mu_{L} e_{ij}^{C}$

"   "   "   "   " inclusion is

$$\sigma_{ij}^{C} - \sigma_{ij}^{T} = \lambda_{L}(e^{C}-e^{T})\delta_{ij} + 2\mu_{L}(e_{ij}^{C}-e_{ij}^{T}) \qquad (4.1)$$

In principle, this method can be used for an arbitrary boundary geometry by integrating the known solution for a point force in an infinite body. For an ellipsoidal boundary the following expressions are given.

$$e_{il}^{C} = S_{ilmn} e_{mn}^{T} \quad \text{in the inclusion.} \qquad (4.2)$$

Some of the coefficients S are zero, while elliptical integrals and the variables ν, a, b and c , where ν is the Poissons' ratio and a, b and c are the principal axes of the ellipsoid, are included in the remaining coefficients. The strain field e_{il}^{C} in the inclusion is homogeneous, whilst in the matrix, it is inhomogeneous.

If the strain field e_{ij}^{C} in the matrix and inclusion is superposed with a homogeneous strain field e_{ij}^{A} , (corresponding stress field: σ_{ij}^{A}) the following fields are obtained

Matrix: Strain field (inhomogeneous): $e^C_{ij} + e^A_{ij}$

Stress   "   ("   ):

$$\sigma^C_{ij} + \sigma^A_{ij} = \lambda_L(e^C+e^A)\delta_{ij} + 2\mu_L(e^C_{ij}+e^A_{ij})$$

Inclusion: Strain field (homogeneous): $e^C_{ij} + e^A_{ij}$

Stress   " (homogeneous):

$$\sigma^C_{ij} + \sigma^A_{ij} - \sigma^T_{ij} = \lambda_L(e^C+e^A-e^T)\delta_{ij}+2\mu_L(e^C_{ij}+e^A_{ij}-e^T_{ij})$$

Imagine that an inclusion of another elastic material, characterized by λ_{1L} and μ_{1L}, is produced. In its unloaded condition, this inclusion is identical in shape to the homogeneous inclusion in its unloaded conditions and can thus replace this without altering the initial field. This heterogeneous inclusion is now imagined deformed with the homogeneous strain field $e^C_{ij} + e^A_{ij}$. If this gives rise to a stress field $\sigma^C_{ij} + \sigma^A_{ij} - \sigma^T_{ij}$ in the inclusion, then this can also replace the homogeneous inclusion in the final field, since the boundary conditions are the same.

The stress-strain field in a heterogeneous inclusion placed in a matrix loaded by σ^A_{ij} (homogeneous far from the inclusion) is thus solved. The conditions to be fulfilled are as follows:

$$\lambda_L(e^C+e^A-e^T)\delta_{ij}+2\mu_L(e^C_{ij}+e^A_{ij}-e^T_{ij})=\lambda_{1L}(e^C+e^A)\delta_{ij}+2\mu_{1L}(e^C_{ij}+e^A_{ij})$$

or

$$\left.\begin{array}{l}\lambda_L(e^C+e^A-e^T)+2\mu_L(e^C_{ij}+e^A_{ij}-e^T_{ij})=\lambda_{1L}(e^C+e^A)+2\mu_{1L}(e^C_{ij}+e^A_{ij}) \\ \qquad\qquad\qquad\qquad\qquad\qquad\qquad\qquad\qquad\qquad\qquad\qquad i=j \\ \text{and} \\ \qquad\qquad 2\mu_L(e^C_{ij}+e^A_{ij}-e^T_{ij})=2\mu_{1L}(e^C_{ij}+e^A_{ij}) \qquad i\neq j\end{array}\right\}\qquad(4.3)$$

Normally, the Lamé constants and e^A_{ij} are known, such that the only unknowns are e^T_{ij}, which can be found from the above-mentioned set of equations. By inserting e^C_{ij}, expressed by e^T_{mn}

$$(\lambda_L-\lambda_{1L})S_{kkmn}e^T_{mn} + 2(\mu_L-\mu_{1L})S_{ijmn}e^T_{mn} - \lambda_L e^T - 2\mu_L e^T_{ij} =$$

$$(\lambda_{1L}-\lambda_L)e^A + 2(\mu_{1L}-\mu_L)e^A_{ij} \quad i = j \qquad\qquad (4.4)$$

and $\quad (\mu_L-\mu_{1L})S_{ijmn}e^T_{mn} - \mu_L e^T_{ij} = (\mu_{1L}-\mu_L)e^A_{ij} \qquad i \neq j$

is obtained.

The coefficients S_{ijmn} $(i \neq j)$ are all equal to zero, except S_{1212} $= S_{1221}$, $S_{1313} = S_{1331}$, and $S_{2323} = S_{2332}$. This means that the shearing strains are given explicitly as

$$e^T_{12} = \frac{\mu_L-\mu_{1L}}{2(\mu_{1L}-\mu_L)S_{1212}+\mu_L} \; e^A_{12} \qquad\qquad (4.5)$$

and the corresponding expressions for e^T_{13} and e^T_{23}.

The coefficients S_{ijmn} $(i = j)$ are zero for $m \neq n$, such that only e_{11}, e_{22} and e_{33} enter the other set of equations. There are thus 3 equations with three unknown values. The first of these equations has the following form:

$$\{(\lambda_L-\lambda_{1L})[S_{1111}+S_{2211}+S_{3311}]+2(\mu_L-\mu_{1L})S_{1111}-\lambda_L-2\mu_L\} \; e^T_{11} +$$

$$\{(\lambda_L-\lambda_{1L})[S_{1122}+S_{2222}+S_{3322}]+2(\mu_L-\mu_{1L})S_{1122}-\lambda_L \quad\} \; e^T_{22} +$$

$$\{(\lambda_L-\lambda_{1L})[S_{1133}+S_{2233}+S_{3333}]+2(\mu_L-\mu_{1L})S_{1133}-\lambda_L \quad\} \; e^T_{33} =$$

$$[(\lambda_{1L}-\lambda_L)+2(\mu_{1L}-\mu_L)]e^A_{11}+(\lambda_{1L}-\lambda_L)e^A_{22}+(\lambda_{1L}-\lambda_L)\;e^A_{33}$$

where S_{ijkl} are complicated expressions. S_{1111}, for example, is expressed by:

$$S_{1111} = S_{2222} = \frac{1}{8(1-\alpha^2)(1-\nu)} \; [\frac{1}{\pi} \; I_a((1-\alpha^2)(1-2\nu) + \frac{9}{4})-3\alpha^2]$$

When $a = b > c$ we have

$$I_a = \frac{2\pi\alpha}{(1-\alpha^2)^{3/2}} \ [\arccos \ \alpha-\alpha(1-\alpha^2)^{\frac{1}{2}}]$$

<u>Homogeneous inclusion of finite stiffness</u>. Numerical solution of the expressions (4.4) for an axially symmetrical ellipsoid, with x_3 as the axis of symmetry, has been carried out in Askegaard.[16,17]. Some of the results are shown in fig. 4.1 and 4.2, where $\beta = \frac{E_1}{E}$. The results correspond to σ_{33} in the inclusion being imagined used to measure σ_{33}^A in the matrix for small values of α (a flat ellipsoid) while e_{33} is imagined used to measure e_{33}^A when α is large. Results corresponding to the following loading cases are shown in the figures:

$$(\sigma_{11}^A, \ \sigma_{22}^A, \ \sigma_{33}^A) = \begin{cases} (0, \ 0, \ \sigma) \\ (\sigma, \ 0, \ 0) \end{cases} \Bigg\} \ \alpha < 1$$

$$(\sigma_{12}^A, \ \sigma_{13}^A, \ \sigma_{23}^A) = (\sigma, \ \sigma, \ \sigma) \Bigg\}$$

$$(e_{11}^A, \ e_{22}^A, \ e_{33}^A) = \begin{cases} (e, \ 0, \ 0) \\ (0, \ 0, \ e) \end{cases} \Bigg\} \ \alpha > 1$$

$$(e_{12}^A, \ e_{13}^A, \ e_{23}^A) = (e, \ e, \ e) \Bigg\}$$

It can be seen that $\sigma_{33} \rightarrow \sigma_{33}^A$ and $\sigma_{ij} \rightarrow \sigma_{ij}^A$ ($i \neq j$) when $\alpha \rightarrow 0$, i.e., the requirement that the inclusion may not change the field in the matrix can be satisfied arbitrarily well, when a small boundary zone can be disregarded, by letting $\alpha \rightarrow 0$. Furthermore, going to the same limit is seen to reduce the dependence of E (the same applies to ν) and dependence of other stress components, i.e., σ_{11} and σ_{22} , to an arbitrarily small value. Conditions 1., 3. and 4. in section 4.1 can thus be satisfied.

The results in the figures are only shown for a single combination

of ν and ν_1. More detailed information can be found in reference 17.

The field in the inclusion for an arbitrary load σ_{ij}^{A} can be found

by superposing the given results.

The problem thus solved can be extended to cover the situation in

which an ellipsoidal inclusion (μ_{1L} and λ_{1L}) placed in an unloaded

matrix (λ_L and μ_L) would attain a homogeneous stress free strain $e_{ij}^{T_1}$

if it could expand freely. This situation arises in, for instance, the case

of shrinkage in the matrix or in the case of alteration of temperature if

the matrix and inclusion have different coefficients of thermal expansion.

The same boundary strain and boundary stress field in such an inclusion

will be obtained as for the homogeneous inclusion described by (4.1) if

the boundary stresses $\sigma_{ij}^{C} - \sigma_{ij}^{T}$ lead the inclusion

(λ_{1L}, μ_{1L}) from the strain field $e_{ij}^{T_1}$ to e_{ij}^{C}, i.e., if

$$\sigma_{ij}^{C} - \sigma_{ij}^{T} = \lambda_{1L}(e_{mm}^{C} - e_{mm}^{T_1})\delta_{ij} + 2\mu_{1L}(e_{ij}^{C} - e_{ij}^{T_1}) \tag{4.6}$$

Combining (4.1) and (4.6) and inserting (4.2) we obtain

$$\left.\begin{array}{l} (\lambda_L - \lambda_{1L})S_{11kk}e_{kk}^{T} + 2(\mu_L - \mu_{1L})S_{ijmn}e_{mn}^{T} - \lambda_L e_{rr} \\[2mm] + \lambda_{1L}e_{rr}^{T_1} - 2\mu_L e_{ij}^{T} + 2\mu_{1L}e_{ij}^{T_1} = 0 \qquad\qquad i = j \\[2mm] 2(\mu_L - \mu_{1L})S_{ijmn}e_{mn}^{T} - 2\mu_L e_{ij}^{T} + 2\mu_{1L}e_{ij}^{T_1} = 0 \qquad i \neq j \end{array}\right\} \tag{4.7}$$

From the latter set of equations, e_{ij}^{T} ($i \neq j$) can be found explicit-

ly and e_{ij}^{T} ($i = j$) can be determined by solving 3 equations with 3 un-

known quantities, if $e_{ij}^{T_1}$ is known. The strain field in the inclusion,

e_{ij}^{C} can then be derived from (4.2). Lastly, the stress field in the inclu-

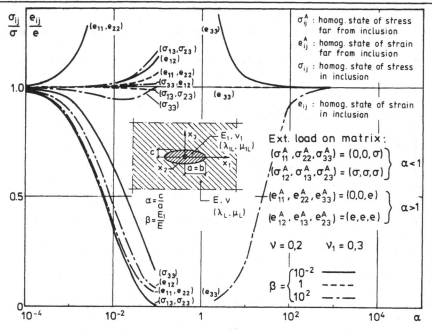

Fig. 4.1   State of Stress and Strain in inclusion when matrix is loaded
           externally.

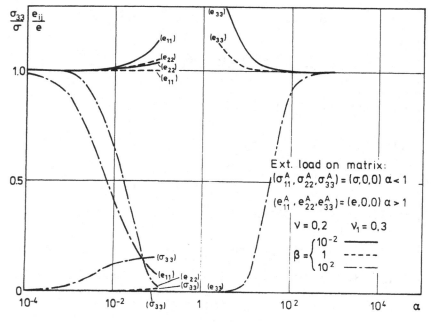

Fig. 4.2   State of Stress and Strain in inclusion when matrix is loaded
           externally.

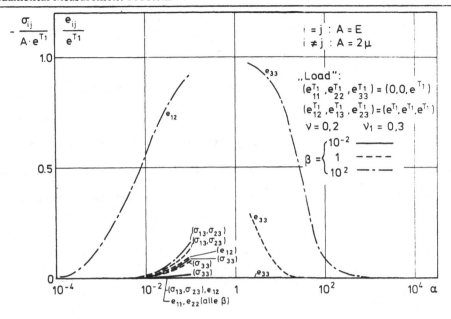

Fig. 4.3   State of Stress and Strain in inclusion from a hindered stress-free
           strain in the inclusion.

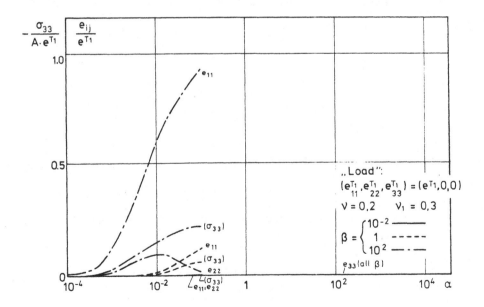

Fig. 4.4   State of Stress and Strain in inclusion from a hindered stress-free
           strain in the inclusion.

sion $\sigma_{ij}^{C} - \sigma_{ij}^{T}$ can be found from the strain e_{ij}^{C} and e_{ij}^{T} by using Hookes' law.

Some of these expressions are also solved numerically. Some of the results can be seen in fig. 4.3 and 4.4. It is now apparent that the influence of a homogeneous, strain-free stress field in the inclusion (or in the matrix) can be made arbitrarily small by letting $\alpha \Rightarrow 0$, i.e., condition 2. in section 4.1 can be satisfied. It is thus now shown that all the conditions can be fulfilled by making α small enough.

It is thus also shown as probable that the stress in a matrix with more complicated constitutive equations can be measured by using inclusions if the stress cell can be constructed such that α is very small. Measurements with pressure cells confirm this.

Similar arguments hold for strain cells when α is very large. See ref. 17.

5. Check of Validity of Assumptions

In chapter 2, the assumptions on which the sets of equations are based are accounted for. Use of the sets of equations for solving practical problems depends on whether observations confirm the validity of the assumptions.

The following are some of the most important of the present assumptions.

a) Assumption of continuum.

b)   Assumption of homogeneity.

c)   Assumption of isotropy.

d)   The initial state is known.

e)   There are constitutive relations between σ_{ij}, e_{ij}, t and θ
     that characterize a material.

f)   The boundary conditions are known. The boundary conditions include
     information about geometry, mechanical- and thermal load, and
     contact with surface-active media.

In the following, these assumptions will be discussed with special
reference to the possibility of testing their validity.

5.1 Assumptions of continuum, homogeneity and isotropy   Homogeneity over a volume in a medium means that there are the same properties at
all points within the volume. This conception of a medium characterizes
a continuum and contradicts the physical observation of an atomic structure.

The assumption of isotropy, too, implies that the medium is a continuum since isotropy means that the properties are the same in all directions at any point.

That the continuum theory can anyway be used in so many cases is due
to the fact that it is often sufficient to determine material properties,
strains, etc. as average values over dimensions (volumes, areas, lengths)
that are large in relation to atomic distances, so that the local variations that exist over the distance between two atoms are smoothed out.

The dimensions over which average values are taken can often be so large that materials containing heterogeneities, e.g. crystals, micro-cracks or particles, can also be assumed to be homogeneous. However, this is conditional upon the heterogeneities being uniformly distributed. This means that such materials as metals, plastic, glass, concrete, wood and granular materials can, in most cases, be assumed to be homogeneous.

This is an advantage, since it means that only one set of constitutive relations has to be set up. At the same time, the only boundary that arises will be the external boundary of the structure. The particular conditions relating to the individual grains or crystals only influence the form of the constitutive equations and the failure criteria.

If each individual grain or crystal were to be regarded as a continuum, this would presumably mean setting up more sets of constitutive equations, because grains with different properties would often be present in the material. More important, however, is the fact that boundary conditions would have to be set up for each grain. This last condition would make calculation of the stress-strain-temperature field in the structure impossible.

However, the last method is of considerable importance to material research and is thus indirectly important to structural research, because knowledge of the stress-strain field at a microscopic level, i.e. in the individual grains, is essential if we are to achieve a more exact understanding of for instance failure mechanisms.

A visual observation will often give a first idea of the length (area, volume) over which a material can be regarded as homogeneous in respect of a given property (e.g. specific weight - a scalar - or stiffness - a tensor).

Determination of the above-mentioned length (area, volume) requires a method for measuring the property that has a better accuracy than the accuracy on the property that is regarded as acceptable for the material to be termed homogeneous. The relationship between the measuring length (area, volume) and the accuracy on the property must be known before the accuracy of the results of a structural measurement can be assessed.

If isotropy is to be investigated, identical measurements must be carried out in different directions, and the relationship between the accuracy on such a set of measurements and the measuring length (area, volume) must be determined.

Example   Fig. 5.1 shows the results of an experiment to determine the measuring length over which concrete can be characterized as homogeneous in respect of a given property.

The desired property here is the stiffness, characterized by the modulus of elasticity $E = \dfrac{\sigma_{11}}{e_{11}}$ (uniaxial stress). (If time dependence is to be taken into account, the same expression can be used, provided the measurements over the different measuring lengths refer to the same time).

A test specimen is used that has such large cross sectional dimensions in relation to the dimensions of the heterogeneities that, from one section to another, the variation of the mean value of e_{11} over a cross section is judged to be negligible. This can be checked by measurement. σ_{11} is taken as the mean stress over the cross section and is known, since the external load is given, and the mean value e_{11} over different measuring lengths of the surface is measured.

The figure shows the results of the two experiments, Askegaard.[18] and

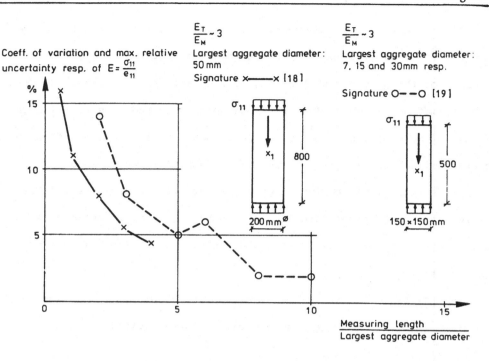

Fig. 5.1 Homogeneity of concrete as function of measuring length-aggregate
          diameter ratio.

Müller.[19] with the coefficient of variation and the maximum relative error

on E depicted as functions of the ratio between measuring length and

largest stone diameter. (The mean value of E can be assumed to be inde-

pendent of the measuring length).

The results are expected to be affected especially by the ratio be-

tween the stiffness of the aggregate E_T and that of the mortar E_M and

by whether the specimen's surface is ground or not. The more the ratio

$\frac{E_T}{E_M}$ deviates from 1, the greater will be the coefficient of variation. Mea-

surement on a ground surface is also expected to increase the coefficient

of variation, since this surface passes both stone and mortar, as opposed
to the untreated surface, which only passes mortar.

The length over which the material can be considered as homogeneous
as regards stiffness is seen to depend on the magnitude of the coefficient
of variation considered as acceptable.

Failure stress (determined under certain external conditions) is another material property for which the material can be examined for homogeneity and isotropy. This leads to a relationship between coefficient of
variation and the ratio between the measuring area and the cross-sectional
area of the largest stone.

The specific weight is a scalar and can thus only be used for testing
homogeneity. Here, we get a relationship between the coefficient of variation and the ratio between the measuring volume and the volume of the largest stone.

5.2 Assumption about a known initial field   The initial field is
of considerable importance for the size of the load that can be imposed
on a structure before its carrying capacity is exhausted. This initial
field is often unknown, e.g., in structures built of elements with an unknown loading history.

As indicated in section 4, the residual stress field can, in some
cases, be determined in castable or granular materials if pressure cells
can be placed at the time of casting or at the time of placing the grains.

Determination of the residual stress field is likewise possible if
the material is linear elastic, isotropic (even anisotropic) and homoge-

nous, with known material constants. In this case, the residual stress is determined measuring the change in the strain field caused by drilling, indentation or sectioning.

There are also certain non-destructive methods with limited uses: x-ray techniques can be used to determine the surface stresses in polycrystalline materials; photo elasticity can be used to determine residual stress fields in glass and certain types of plastic. The ultra-sound method can also be used in some cases, since the propagation volocity of a transverse wave is dependent of the stress field in a plane perpendicular to the direction of propagation of the wave.

5.3 <u>Experiments for determination of the constitutive equations,</u> <u>notably the $\sigma_{ij} - e_{ij} - t - \theta$ relationship</u>   Experiments to determine constitutive equations are planned to simplify interpretation of their results, i.e. the specimen is designed and loaded (thermally or mechanically) such that one or more quantities in the equations are known and such that the remaining quantities can be measured.

This demand for simplicity leads to the use of specimens with a homogeneous field. The field of the specimen is only homogeneous if the initial field is homogeneous, i.e. if the specimen is free from residual stresses or the moisture content is uniformly distributed etc. The initial field must therefore be checked.

Surface roughness may also give deviations from a homogeneous state in a test specimen due to local stress concentrations. This influences, in particular, the rupture load for brittle materials. It should be borne in mind that fatique loaded metals also behaves as brittle materials.

Determination of the relationship between the parameters in the constitutive equations is considerably more difficult in the case of time-
dependence, anisotropy and non-linearity, singly or concurrently.

An attempt to describe a material with unknown properties will presumably be based on the results from an experiment in which a uniaxial field
has been produced in the specimen. Provided that the stability and sensitivity of the measuring equipment is good enough, this will reveal the
existence of time-dependence.

If the material can be assumed to have time-independent properties
in some applications, it will be necessary next, to investigate whether
there is an interval in which there is a linear relationship between the
parameters. If the relationship is non-linear, it must, if possible, be
approximated by a suitable mathematical function. If the material has time-
dependent properties, it will be necessary to investigate whether the relationship can be described by simple mathematical expressions, e.g. if
the $\sigma_{ij} - e_{ij} - t$ relationship is under examination, whether the material can be characterized as linear, visco elastic.

If the material can be described in a uniaxial field, generalization
to a triaxial field with subsequent control tests should be attempted, unless the structure on which the results are to be used is composed only of
elements with a uniaxial field.

If generalization of the constitutive equations is successful, it
should, in principle, be possible to predict the field in any problem.
However, if the field is complicated (complicated geometry or loading), the

calculation may be impossible even if the constitutive equations are fully determined. In such cases, information must be obtained from direct experiments with the structural element or with a specimen where the same situation has been approximated.

Experiments of this nature are called technical control experiments.

Shape and loading of the specimen. In the design of test arrangements it is essential to avoid eccentricities in the loading and to ensure that the stress fields in the failure zone is homogeneous and known right through to failure.

If it proves impossible to avoid eccentricities, the field of the specimen will not be homogeneous. In such case, the magnitude of the eccentricity should be determined and the strain measured at several points of the circumference of the prism so that the results can be corrected. In general, a check should be made for eccentricities in each experiment.

A few test arrangements with a homogeneous uniaxial field at varying load intervals, - in some cases right through to failure - are shown in figures 5.2 to 5.6.

It is envisaged that the control tests are carried out at constant temperature, and they should be repeated at other temperature levels so that the materials' properties and applicability can be determined.

Methods for specimens with an inhomogeneous field in the measuring area are called technical control methods. This category includes tests with beams and columns in which there is an unknown inhomogeneous initial field. Examples of technical control tests are shown in fig. 5.7 - 5.10.

Technical control tests are important, but have their limitation in

EXPERIMENTAL ARRANGEMENTS WITH HOMOGENEOUS STRESS STATE IN THE MEASURING SECTION.				
Figure	Stress state	Material	Test set up	Comment
5.2	com-pres-sion	arbitrary solid	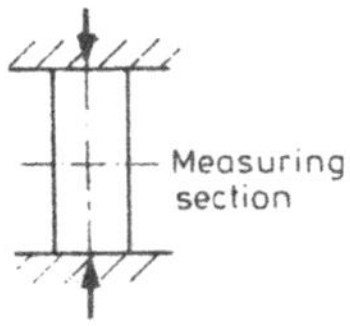 Measuring section	The existence of friction at the contact surfaces means that crack-pattern at rupture and rupture load depend on the height-diameter ratio. When the ratio is > 3, the field will be homogeneous at the measuring section.
5.3	com-pres-sion	"	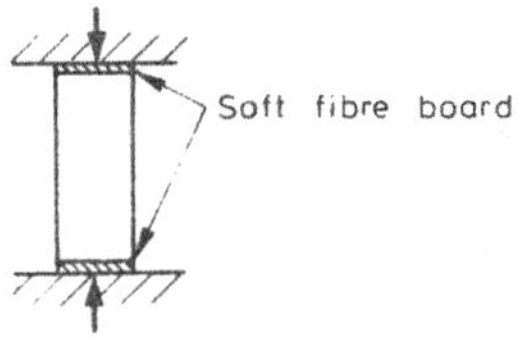 Soft fibre board	As for fig. 5.2 Reduced dependence of height-diameter ratio.
5.4	com-pres-sion	"	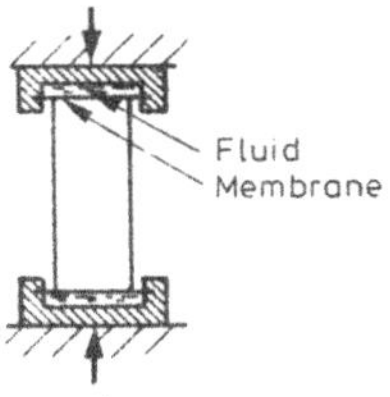 Fluid Membrane	Homogeneous field everywhere in the test specimen. Max. load depends on the behaviour of the rubber membrane. Gravesen, Krenchel.[20] have obtained 12.5 MPa
5.5	ten-sion	concrete	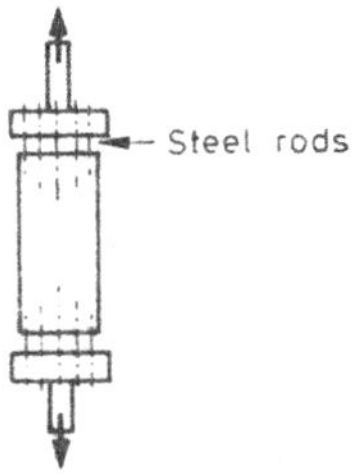 Steel rods	Stress concentrations around the steel rods casted in the concrete are about the same as around the aggregates. Homogeneity up to rupture can therefore be obtained, see Gravesen, Krenchel.[20] .
5.6	ten-sion	brittle	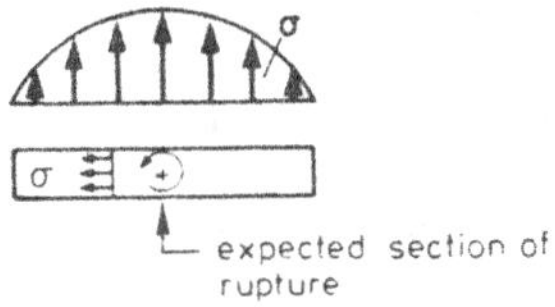 σ expected section of rupture	The prismatic test specimen rotates about an axis through the center of gravity. The stress field can be considered homogeneous up to rupture at the mid-section. Berthier.[21], Askegaard.[22]

The stress state column also notes: uniaxial static or quasistatic

EXPERIMENTAL ARRANGEMENTS WITH HETEROGENEOUS STRESS STATE (technical control tests)			
Fig-ure	Material	Test set up	Comment
5.7	Soft and hard plas-tics. Metal of low stiffness.		Torsional eigen-frequency of vi-brating system is used. Depen-dence of stiffness and damping on frequency is investigated in the interval 0.01-50 Hz.
5.8	Reinforced concrete- and wooden beams. Steel profiles.		Test is carried out with the structural member itself under realistic conditions. Initial field (residual stress field) is unknown.
5.9	Reinforced concrete.		Pulling out test of reinforcing bar. Initial field is not well known.
5.10	Metals.		Information is obtained about fracture toughness of the mate-rial from impact tests.

the fact that the results are difficult to use for predicting the field in specimens with a slightly different geometry, or with other loads.

Determination of the material constants for linear elastic, orthotrop-ic material. In the following an orthotropic material will be used as an example of how to determine material constants. It is assumed that the ma-terial has been proved by experiment to have time-independent properties and, further that a loading interval in which there is a linear relation

between stresses and strains has been proved to exist. Tests are only per-

formed in cases in which the strains are small. Furthermore, it is assumed

that the tests are carried out at a constant temperature.

In addition it is assumed possible to determine, by observation,

whether the material is orthotropic, i.e., with 3 planes at right-angles

to one another, indicating structural symmetry. Some wood specimens and cer-

tain fibre-reinforced materials are examples of such materials.

By using the structural symmetry, C_{ijkl} in (2.9) reduces to 9 inde-

pendent material constants, see, for example, Sokolnikoff.[23], Hearmon.[24].

If a cartesian coordinate system with axes in the symmetry planes is used,

then the generalized Hookes' law can be expressed by (5.1)

$$
\begin{bmatrix} e_{11} \\ e_{22} \\ e_{33} \\ e_{23} \\ e_{13} \\ e_{12} \end{bmatrix}
=
\begin{bmatrix}
s_{11} & s_{12} & s_{13} & 0 & 0 & 0 \\
s_{12} & s_{22} & s_{23} & 0 & 0 & 0 \\
s_{13} & s_{23} & s_{33} & 0 & 0 & 0 \\
0 & 0 & 0 & s_{44} & 0 & 0 \\
0 & 0 & 0 & 0 & s_{55} & 0 \\
0 & 0 & 0 & 0 & 0 & s_{66}
\end{bmatrix}
\begin{bmatrix} \sigma_{11} \\ \sigma_{22} \\ \sigma_{33} \\ \sigma_{23} \\ \sigma_{13} \\ \sigma_{12} \end{bmatrix}
\qquad (5.1)
$$

If the coordinate system is rotated in relation to the symmetry planes,

the number of material constants will remain unaltered, but the material

constant matrix in (5.1) will become considerably more complicated. The

same applies to interpretation of the results of the tests, since it must

be presumed that none of the material constants can be determined expli-

citly, but must be determined by solving a set of equations in which they

occur as unknown quantities.

If a coordinate system is chosen such that it leads to the matrix in
(5.1), it will be seen that the 9 material constants can be determined
from 6 experiments performed on specimens cut out in different directions
from the body and loaded as shown in fig. 5.11. The field in the measuring
area is homogeneous in all tests. However, owing to special test problems
specimen with an inhomogeneous field may be more attractive. For acceptance of an inhomogeneous field in a specimen the material must be linear-
elastic. This condition is satisfied here. Tests of this type are described
in Hearmon.[24] .

If the x_1-x_2 plane and all planes perpendicular to this are planes of
structural symmetry the number of material constants reduces to 5 as given
in expression (5.2). Such a material is called transverse isotropic.

$$
\begin{bmatrix} e_{11} \\ e_{22} \\ e_{33} \\ e_{23} \\ e_{13} \\ e_{12} \end{bmatrix}
=
\begin{bmatrix}
s_{11} & s_{12} & s_{13} & 0 & 0 & 0 \\
s_{12} & s_{11} & s_{13} & 0 & 0 & 0 \\
s_{13} & s_{13} & s_{33} & 0 & 0 & 0 \\
0 & 0 & 0 & s_{44} & 0 & 0 \\
0 & 0 & 0 & 0 & s_{44} & 0 \\
0 & 0 & 0 & 0 & 0 & (s_{11}-s_{12})
\end{bmatrix}
\begin{bmatrix} \sigma_{11} \\ \sigma_{22} \\ \sigma_{33} \\ \sigma_{23} \\ \sigma_{13} \\ \sigma_{12} \end{bmatrix}
\qquad (5.2)
$$

In the isotropic case there are 2 constants as given in (5.3)

$$
\begin{bmatrix} e_{11} \\ e_{22} \\ e_{33} \\ e_{23} \\ e_{13} \\ e_{12} \end{bmatrix}
=
\begin{bmatrix}
s_{11} & s_{12} & s_{12} & 0 & 0 & 0 \\
s_{12} & s_{11} & s_{12} & 0 & 0 & 0 \\
s_{12} & s_{12} & s_{11} & 0 & 0 & 0 \\
0 & 0 & 0 & (s_{11}-s_{12}) & 0 & 0 \\
0 & 0 & 0 & 0 & (s_{11}-s_{12}) & 0 \\
0 & 0 & 0 & 0 & 0 & (s_{11}-s_{12})
\end{bmatrix}
\begin{bmatrix} \sigma_{11} \\ \sigma_{22} \\ \sigma_{33} \\ \sigma_{23} \\ \sigma_{13} \\ \sigma_{12} \end{bmatrix}
\qquad (5.3)
$$

LINEAR ELASTIC, ORTHOTROPIC MATERIAL		
Fig. 5.11	x_3, x_2 Planes of symmetry x_1x_2, x_2x_3, x_1x_3.	
Test	Measured quantity	Material constant determined
a	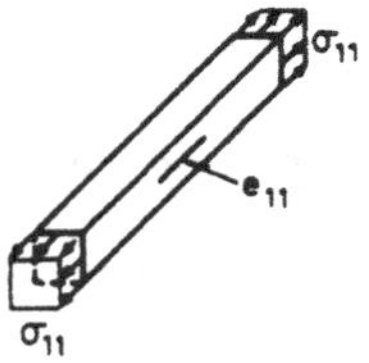 e_{11}	s_{11}
b	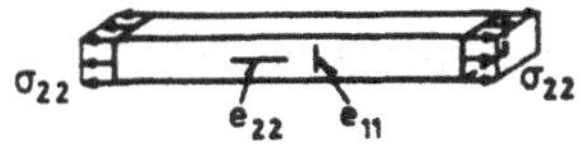 $e_{22}\ e_{11}$	$s_{22}\ s_{12}$
c	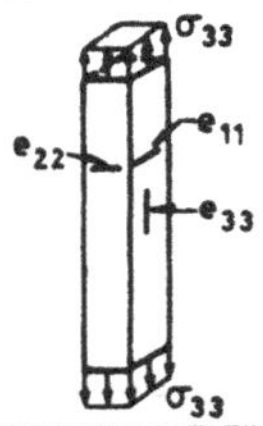 $e_{33}\ e_{11}\ e_{22}$	$s_{33}\ s_{13}\ s_{23}$
d	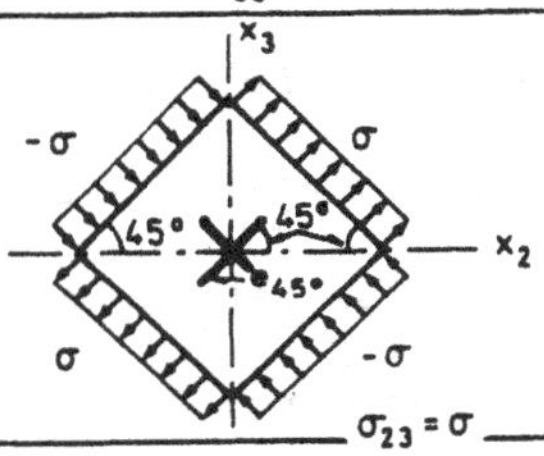 $e_{23} = \dfrac{e_{45°} - e_{-45°}}{2}$	s_{44}
e	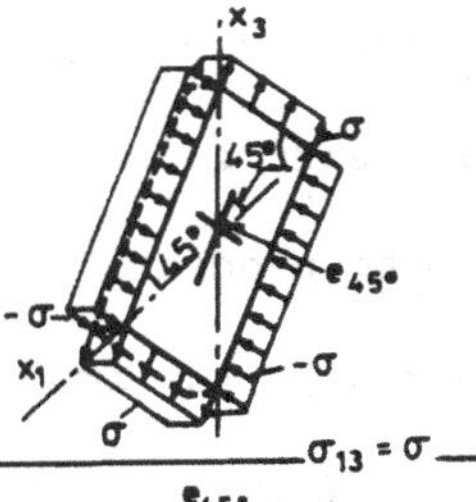 $e_{13} = \dfrac{e_{45°} - e_{-45°}}{2}$	s_{55}
f	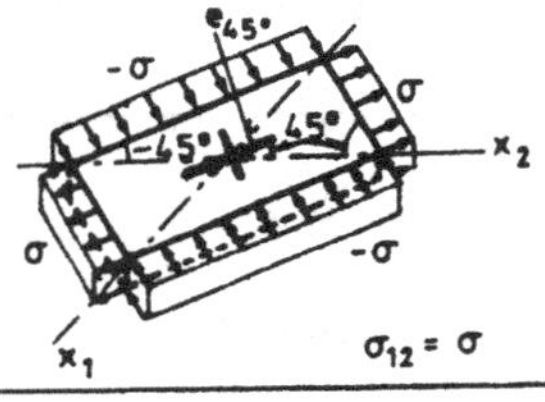 $e_{12} = \dfrac{e_{45°} - e_{-45°}}{2}$	s_{66}

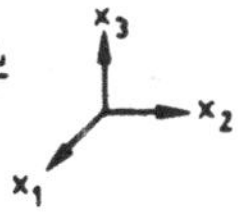

If $s_{11} = \frac{1}{E}$ and $s_{12} = \frac{\nu}{E}$ is inserted, (5.3) can be written as in (5.4).

$$e_{ij} = -\frac{\nu}{E}\sigma_{kk}\delta_{ij} + \frac{1+\nu}{E}\sigma_{ij} \tag{5.4}$$

Determination of the material functions for linear visco-elastic, isotropic materials. As given in section 2.4, a linear, visco-elastic material is characterized by 2 time-dependent functions. If $E(t)$ and $\nu(t)$ are used as material functions, the constitutive equations will have the form (5.5)

$$e_{ij} = \int_{0}^{t}\left[-\frac{\nu(t-\tau)\delta_{ij}}{E(t-\tau)}\frac{\partial\sigma_{kk}(\tau)}{\partial\tau} + \frac{1+\nu(t-\tau)}{E(t-\tau)}\frac{\partial\sigma_{ij}(\tau)}{\partial\tau}\right]d\tau \tag{5.5}$$

If the functions $E(t)$ and $\nu(t)$ are known in the interval $0 < t < \infty$, the strain caused by a known, time-dependent, stress can be predicted. An experiment often used to determine the 2 functions is the creep test with a uniaxial stress field applied at $t = 0$ and thereafter kept constant. This test does not offer a very precise determination of the 2 functions for the very small values of t that are, for example, important, when a visco-elastic structure is subjected to dynamic loading.

Other time-dependent loading is therefore necessary if the field in a structure in which phenomena occur rapidly is to be predicted.

At this point, special attention should be drawn to the important test in which the specimen is loaded with a harmonic force which induces a uniaxial stress field in the specimen. This test results in the material being described by means of other timedependent functions. Only if the functions are known throughout the whole frequency interval $0 < f < \infty$, or in a considerable part of this frequency range, can one of the sets of material functions be found from the other, see Ferry.[10]. This transition from one set of material functions to another is important, if we are to be able to compare the results of experiments carried out in different ways.

A specimen as shown in fig. 5.12 is loaded with a harmonic force of such frequency that the stress field in the specimen is uniaxially homogeneous $\sigma_{11}(t)$, i.e. such that the time it takes a stress wave to run along the specimen is short compared with the period of the external force and such that the inertia forces do not alter the stress field.

The harmonic loading inserted in (5.5) leads to a harmonic strain with the same period, but with a phase-shift in relation to loading.

The stress-strain relationship can be written

$$e_{11} = \sigma_o (D_1 \cos \omega t + D_2 \sin \omega t) \tag{5.6}$$

or

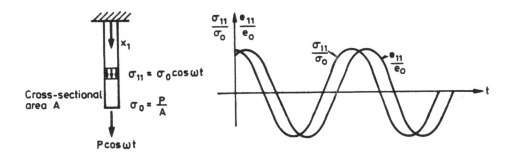

Fig. 5.12 Harmonically loaded linear visco-elastic specimen

$$\left.\begin{array}{l} e_{11} = e_o \cos(\omega t - \delta) \\[2mm] \text{where} \\[1mm] \quad e_o = \sigma_o \sqrt{D_1^2 + D_2^2} \\[2mm] \text{and} \\[1mm] \quad \tan \delta = \dfrac{D_2}{D_1} \end{array}\right\} \tag{5.7}$$

D_1 and D_2 are frequency-dependent but not time-dependent material functions.

The concept complex material functions is often introduced. If σ_{11} and e_{11} are regarded as the real part of a complex stress $\sigma_{11}^* e^{i\omega t}$ and a complex strain $e_{11}^* e^{i\omega t}$, respectively where σ_{11}^* and e_{11}^* are complex quantities and e is the exponential function, then the complex modulus of elasticity E^* is defined as the ratio between σ_{11}^* and e_{11}^*. The following is thus valid:

$$\left.\begin{array}{l} E^* = E_1 + iE_2 = \dfrac{\sigma_{11}^*}{e_{11}^*} = \dfrac{\sigma_o}{e_o} e^{i\delta} \\[3mm] \text{where} \\[2mm] \dfrac{\sigma_o}{e_o} = \dfrac{1}{\sqrt{D_1^2 + D_2^2}} \quad \text{and} \quad \tan \delta = \dfrac{D_2}{D_1} \end{array}\right\} \tag{5.8}$$

The complex form of the material function is convenient because the relationship between the complex material functions (characterized by the same type of loading, i.e. harmonic at the same frequency) has the same form as between the corresponding material constants for a linear elastic material.

For a linear elastic material, for example, the following relation, exists between the modulus of elasticity E, shear modulus G and Poissons' ratio ν:

$$E = 2G(1 + \nu) \tag{5.9}$$

The corresponding relationship expressed by means of complex material functions is:

$$E^* = 2G^*(1+\nu^*)\tag{5.10}$$

The same simple relationship does not exist between E_1, G_1 and ν_1 or E_2, G_2 and ν_2. The relationship between these can be obtained by solving (5.10), (5.8) and the corresponding expressions for G^* and ν^*.

Well-hardened concrete, wood and plastics are examples of materials considered by some authors to be linear, visco-elastic materials.

5.4 <u>Determination of boundary conditions</u>   The design of a structure is based on assumptions of a specific geometrical shape and a specific mechanical and thermal loading. In addition, an evaluation of the importance of surface contact between the material used in the structure and other materials should be included. This effect is not covered by the thermomechanical continuum theory, which means that it must be taken into consideration either in the constitutive equations, in rupture criteria or as a boundary load which is done in certain hydrodynamic problems.

<u>Boundary geometry</u>. Minor deviations from the assumed geometrical shape are often insignificant, but there are considerable exceptions to this, e.g. where small geometrical deviations cause an alteration of the expected stress field, resulting in a greatly reduced load carrying capacity for the structure (elastic instability).

Visual inspection, photogrammetry or the moiré-technique can be used to check the initial geometry.

Cracks also represent part of a structure's geometry. If the cracks
are uniformly distributed and are very small in relation to the structural
parts of the structure, it is advantageous, as indicated in section 5.1,
to let their existence be expressed in the formulation of the constitutive
equations, i.e. to make allowance for them when determining material properties.

On the other hand, if the cracks are large or not uniformly distributed, they must be considered as a part of the structure's external or internal geometry, unless they are located in such a way that their presence
does not alter the stress field.

The latter is normally considered to be the case with shrinkage cracks
in wooden structural elements in which there is a uniaxial stress field in
the fibre-direction of the wood.

The presence of cracks is of great importance to the loading capacity
of structures built of brittle materials in which tensile stress fields
occur. This is the case, for instance, where steel structures are subjected
to dynamic effects or to multiaxial tensile stress fields.

Such structures, e.g., pressure vessels, must be inspected closely
for cracks. The most common methods for this are the X-ray method, the
ultra-sound method and registration of acoustic emission. The risk of
possible crack growth causing total failure is thereafter evaluated. This
evaluation necessitates the addition of further assumptions to the continuum theory, as the field in the zone around the ends of the fractures
is assumed to be dependent on the atomic structure and is thus not de-

scribed by the continuum theory. This special topic called fracture mechanics is described for instance in Kobayachi.[25], Lawn, Wilshaw.[26] and Liebowitz.[27] . Another procedure in which the continuum concept is maintained,
but which involves deviating from the axiom of neighborhood, (i.e. where
the stress field at one point is also presumed to be dependent of the field
outside the immediate neighborhood of this point), is described by Eringen
and Kim.[28] .

In brickwork and unreinforced concrete, in both of which the material
must be characterized as brittle, with very low tensile strength, the problem of cracking is seldom serious. This is because these materials are only
used in structures, in which compressive stresses dominate, and in which

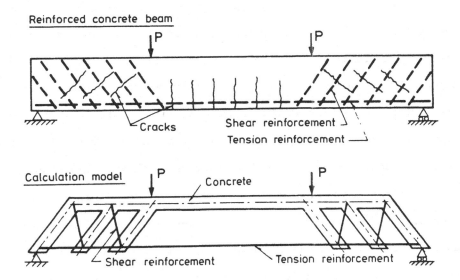

Fig. 5.13 Calculation of the carrying capacity of a reinforced concrete
          beam neglecting the real crack geometry.

crack formation therefore only causes a minor change of the stress field.

In connection with the design of reinforced concrete structures, in which both tensile and compressive stresses exist, crack formation may be considered as an alteration of the geometry of the structure to be described by the equations from fracture mechanics, see e.g. Rostam and Byskov.[29] .

However, the normal procedure is to use the expected fully developed crack distribution to make a simplified analytical model as shown in fig. 5.13.

The procedure described gives good determination of the load capacity for simple structures, but naturally does not describe the geometry of the real cracks.

The same procedure can be used in connection with more complicated reinforced concrete structures. By using the finite element method, various constitutive equations for the concrete can be introduced. The calculated load capacity of these more complicated structures has not been verified by tests to the same extent as the loading capacity of simple structures.

Boundary loading. Structures are often designed for loading cases described in codes. This simplifies calculation as these loading cases constitute a simplification of the real load situation. This procedure is reasonable if the codes are based on the results of extensive tests.

Another procedure that is now gaining ground is one in which the quantities included in the calculation are characterized by a mean value and a standard deviation. Mean value and standard deviation are thus used for quantities characterizing load, material properties, geometry, etc.

The design criterium is that the risk of structural failure must not exceed a certain value. This procedure assumes previous extensive tests, because the distribution functions for the quantities must be determined experimentally.

The methods described above for establishment of a design criterium are discussed in the literature, see e.g. Borges, Castanheta.[30] and Dyrbye, Gravesen, Krenk, Lind, Madsen.[31] .

Attention should also be drawn to the boundary conditions at the supports of the structure, where assumptions about frictionless bearings, complete restraint or similar conditions are often used in calculations and may represent too rough an approximation.

<u>Surface contact with other materials</u>. The energy state of an atom inside a body deviates from the energy state of an atom near the surface of the body.

This effect is not accounted for in the thermo-mechanical continuum theory but is approximated by assuming a surface tension field, i.e. a boundary loading with the dimension force per unit of length exerted on the surface and in its plane. The magnitude of the surface tension is material-dependent, i.e. dependent on both the material inside the body and that surrounding it.

The direct contribution of this effect to the stress and strain field in solids is neglected, because it is insignificant. However, the effect may be of importance in hydrodynamics, when contact faces arise between gases and liquids. The surface stress field is also of importance to the rise in a capillary tube at contact faces between solids, liquids and gases.

Even though direct contribution of the surface contact to the alteration of the stress field in a solid body is negligible, this effect may
have considerable consequences for the constitutive equations and for failure conditions.

The results of creep tests with concrete, using air and water, resp.,
as the contact media are given in fig. 5.14 as an example of alteration of
the constitutive equations. (Hannant.[32]). It is seen that creep in air from
loading of specimen at time 110 days until 420 days is about 80 μstr. By
changing from air to water at the same temperature the creep is increased
drastically. It is about 1050 μstr. in 160 days from time 420 to 580 days.
The creep is corrected for expansion due to water absorption.

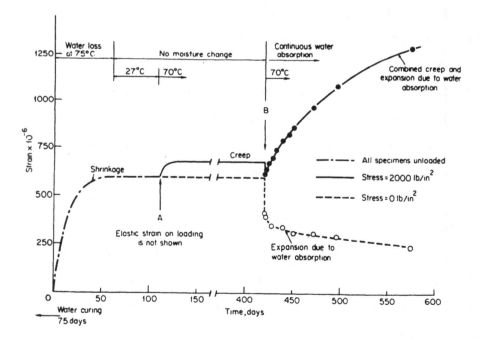

Fig. 5.14   The effect of water movement on the time dependent strains for
concrete at 70°C - 75°C. (Elastic and thermal strains omitted)
(after Hannant.[31])

This shows that homogeneity of the water content is essential in the test specimens used to determine the constitutive equations in section 5.3.

The change of contact medium is identical to an alteration of the work that must be done to produce new surfaces. This is also the case at the tips of cracks, where contact molecules can alter the initial condition so that less work is required to make the cracks grow.

As an example of the dependence on the contact media of the rate of crack propagation fig. 5.15 shows the results of a test with a cracked steel specimen under constant external loading in which hydrogen and a mixture of hydrogen and oxygen are used as contact medium entering the crack. A small amount of oxygen added to the hydrogen is seen to stop the

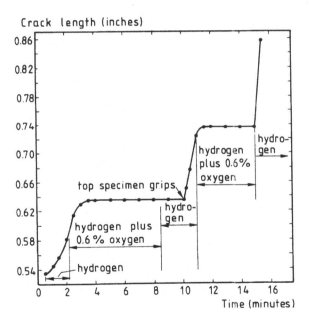

Fig. 5.15 Subcritical crack growth in hydrogen and a hydrogen-oxygen mixture in martensitic high-strength steel. (After Hancock, Johnson.[33]) . © 1966 AIME, New York, N.Y.

crack growth (Hancock, Johnson.[33]).

It is often possible to determine whether there is any microcrack propagation in a solid body, because the short mechanical wave pulses, called acoustic emission, caused by the crack growth can be registered on the surface of the body.

In the above discussion of contact problems chemical decomposition, i.e., effects which also take place, even when the body is mechanically unloaded is neglected.

Readers interested in surface contact problems are referred to e.g. Rehbinder, Shchukin.[34] and Aveyard, Haydon.[35].

6. References

1. Leipholz, H.H.E., Analytical foundations of experimental mechanics, in *Study No. 9: Experimental Mechanics in Research and Development*, Unversity of Waterloo, Waterloo, 1973, 43.

2. Drucker, D.C., Thoughts on the present and future interrelation of theoretical and experimental mechanics, *Proc. S.E.S.A.*, 25, 1, 97, 1968.

3. Pindera, J.T., Contemporary trends in experimental mechanics: Foundations, methods, applications, in *Study No. 9: Experimental Mechanics in Research and Development*, University of Waterloo, Waterloo, 1973, 143.

4. Eringen, A. Cemal, *Mechanics of Continua*, John Wiley & Sons, New York, 1967.

5. Liepmann, H.W. and Roshko, A., *Elements of Gasdynamics*, John Wiley & Sons, New York, 1957.

6. Landau, L.D. and Lifshitz, E.M., *Fluid Mechanics*, Pergamon Press, 1959.

7. Hodge, P.G., *Continuum Mechanics*, McGraw-Hill, New York, 1970.

8. Gurtin, M.E. and Sternberg, E., On the linear theory of viscoelasticity, *Archive Rat. Mech. Analysis*, 11, 1962.

9. Fung, Y.C., *Foundations of Solid Mechanics*, Prentice-Hall, New Jersey, 1965.

10. Ferry, J.D., *Viscoelastic Properties of Polymers*, John Wiley & Sons, New York, 2nd Ed., 1970.

11. Zienkiwicz, O.C., *The Finite Element Method*, 3rd Ed., London, 1977.

12. Stanley, P., (edit.) *Computing Developments in Experimental and Numerical stress analysis*, Appl. Sci. Publishers, London, 1976.

13. Stein, P.K., A unifying approach to experimental mechanics, in *Study No. 9: Experimental Mechanics in Research and Development*, University of Waterloo, Waterloo, 1973, 475.

14. Stein, P.K., *Measurement engineering*, vol. 1: Basic Principles, Stein Engineering Services, Phoenix, Arizona, 1964.

15. Eshelby, J.D., The determination of the elastic field of an ellipsoidal inclusion and related problems, in *Proc. of the Royal Society, Series A,* 241, 1957, 376.

16. Askegaard, V., Stress and Strain measurements in solids, in *Proc. 6th Int. Conf. on Exper. Stress Anal.*, München, 1978, 259.

17. Askegaard, V., Stress and strain measurements in solid materials, *Report R 92, Struct. Res. Lab*. Techn. Univ. of Denmark, 1978.

18. Askegaard, V., Strain measurements on concrete (in Danish), *Proj. 5513, Struct. Res. Lab.*, Techn. Univ. of Denmark.

19. Müller, R.K., Der einfluss der messlänge auf die ergebnisse bei dehnmessungen an beton, *Inst. für Massivbau*, Techn. Hochsch. Darmstadt, 1962.

20. Gravesen, S., Krenchel, H., Compressive-, tensile- and bending strength of concrete (in Danish), *Proj. 1/62, Struct. Res. Lab.*, Techn. Univ. of Denmark.

21. Berthier, R.M., Étude et controle des caractéristiques pratiques des ciments et bétons, *Publication Techn. No. 27, Centre d'étude et de recherches de l'industrie des liants hydrauliques*, Paris, 1950.

22. Askegaard, V., Tensile strength tests on bars made from brittle mate-
    rials, (in Danish), in *"Avancerade metoder för byggnadsmaterialforsk-
    ning och -provning"*, Valtion teknillinen tutkimuslaitos, Helsinki,
    1968, 5.

23. Sokolnikoff, I.S., *Mathematical Theory of Elasticity*, McGraw-Hill,
    1956.

24. Hearmon, R.F.S., *An Introduction to Applied Anisotropic Elasticity*,
    Oxford University Press, Oxford, 1961.

25. Kobayashi, A.S., (edit.) *Experimental Techniques in Fracture Mechanics*,
    Iowa State University Press, Ames, 1973.

26. Lawn, B.R. and Wilshaw, T.R., *Fracture of Brittle Solids*, Cambridge
    University Press, Cambridge, 1975.

27. Liebowitz, H., *Fracture, vol. 1-7*, Academic Press, 1968-72.

28. Eringen, A. Cemal and Kim, B.S., On the problem of crack tip in non-
    local elasticity, in *"Continuum Mechanics Aspects of Geodynamics and
    Rock Fracture Mechanics"*, Thoft-Christensen, P., (edit.) Reidel pub-
    lishing company, Dordrecht, 1974, 107.

29. Rostam, S. and Byskov, E., Fracture mechanics approach to determine
    crack lengths in concrete structures utilizing the finite element
    method, *CEB. Bulletin d'Information No. 89, section 2.7,* 1973, 16.

30. Borges, J. Ferry and Castanheta, M., *Structural Safety*, Laboratorio
    Nacional de Engenharia Civil, Lisbon, 1971.

31. Dyrbye, C., Gravesen, S., Krenk, S., Lind, N.C., Madsen, H.O.,
    Safety of Structures (in Danish), Den Private Ingeniørfond, Techn.
    Univ. of Denmark, Copenhagen, 1979.

32. Hannant, D.J., An experimental investigation of the creep mechanics
     in concrete, in *Proc. Southampton 1969 Civ. Eng. Conf. on Structure,
     Solid Mechanics and Engineering Design,* part 1, Wiley-Interscience,
     331.

33. Hancock, G.G. and Johnson, H.H., Hydrogen, oxygen, and subcritical
     crack growth in a high-strength steel, *Trans. Met. Soc. AIME,* 236,
     513, 1966.

34. Rehbinder, P.A. and Shchukin, E.D., *Surface Phenomena in Solids
     during Deformation and Fracture Processes,* Progress in Surface science,
     3, part 2, 1972.

35. Aveyard, R. and Haydon, D.A., *An Introduction to the Principles of
     Surface Chemistry,* Cambridge University Press, Cambridge, 1973.

ADVANCED THEORETICAL AND EXPERIMENTAL ANALYSIS OF

PLATES AND PLATES IN CONTACT

Prof. Dr.-Ing. K.H. Laermann,
o.Prof. für Baustatik, Fachbereich 11 -
Bautechnik, Universität - GH - Wuppertal,
Pauluskirchstr. 7,
D-5600 Wuppertal 2

1. Introduction

In order to describe any kind of process, i.e. biological, physical, psychological, social and economical processes, "models" are to develop. The results of any investigation, of any analysis, their reliability and their accuracy strongly depend on the fact, that the "model" must describe the reality exactly or as exact as possible in such a way, that the numerous parameters with considerable influence on the phenomena observed in real "life" are included. It is also important to consider the interdepending facts in the regarded systems or subsystems according to the principles of cybernetics.

As it is impossible - or at least as it seems to be - to model the universal "events" comprehensively, we have to start with subsystems, combining them under consideration of interactions and reactions step by step towards larger systems, which we may denote as "macrosystems". The scientific as well as the technological progress enables us more and more to proceed in such a way.

But still limited in our scientific capacity and technical possibilities despite the tremendous progress, for example in computer- and data processing techniques, we will deal in the field of physical processes with the subsystem "mechanical problems".

If the physical process in a real structure is to describe in order to predict e.g. the internal stresses, strains, deformations or the safety of structures against failure, i.e. the reactions to different actions, the structure itself as well as the actions on the structure must be described in a "mathematical model"; operational models must be produced to describe the external and internal events. Generally it is impossible to model the real process, i.e. to derive the "true" constitutive equations or - if this should nevertheless be possible - to solve these constitutive equations. We have to make assumptions and simplifications on very different levels, which only lead to an approach of reality, the certainty and reliability of which may be unknown very often.

However, it seems to be possible to introduce mechanical models to
analyze the problem experimentally much more realistic. But again we need
a mathematical model in order to transmit or to evaluate experimental
data. As in many cases the model material is not the same as the material
of the real structure, the different response of materials must also be
taken into consideration. Proper models of material behaviour must be introduced more or less describing the real behaviour of the materials.
Therefore it must be pointed out that even the experimental analysis
leads to an approximated solution only. But generally, the approach is
much better because more complex. Therefore more realistic mathematical
models can be considered, as the constitutive equations are solved by experiments instead of numerically.

With the development of new experimental techniques, methods and
equipment accuracy and reliability of measured values have become much
higher and are still increasing. New fields of application of experimental
analysis are opened. With respect to data processing and evaluation by
means of computer techniques, it is therefore necessary as well as conse-
quent to derive more complex and reliable theoretical models.

It becomes obvious, that a combination of mathematical and mechanical models, a combination of theory and experiment yields a better knowledge of structure reactions and most reliable results in stress analysis.
In consequence, this leads to "hybrid techniques", the principle of which

is shown in fig. 1, where the different loops are demonstrated.

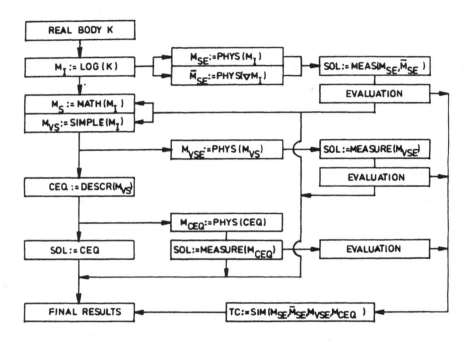

Fig. 1: Block diagram of "hybrid technique"

According to the previously stated principles, plates will be regarded as a micro-subsystem. Large deflections, in-plane stresses, contact to
a yielding subgrade, contact areas, friction stresses in the interface
between plates and subgrade will be considered. An extended plate theory
will be introduced, and some experimental methods are described to take
the encountered parameters into consideration.

2. Mathematical Model of an Extended Plate Theory

Especially in experimental analysis of plate-in-bending problems, but also in practice, the deflection of plates is larger than the thickness, and the classical plate theory does not describe the real stress state. In such cases, an in-plane as well as a bending stress state exists simultaneously. This is also true, when the plate is supported on a yielding subgrade, i.e. for contact problems. Friction stresses in the contact area must be considered as well as the real contact area, as dilamination and uplifting may be possible.

Therefore an extended non-linear plate theory is derived, Wolmir[1], Laermann[2], under the following assumptions:

- The hypothesis of Bernoulli and Hooke's Law of Elasticity are still valid;
- homogeneous and isotropic response of materials;
- neglect of the stresses σ_z perpendicular to the neutral plane of the plate.

According to fig. 2, the in-plane strains can be described by the displacements in the xy-plane

$$\varepsilon'_x = u,_x + \frac{1}{2} u,^2_x + \frac{1}{2} v,^2_x \simeq u,_x \qquad (1.1)$$

$$\varepsilon'_y = v,_y + \frac{1}{2} v,_y^2 + \frac{1}{2} u,_y^2 \simeq v,_y \tag{1.2}$$

$$\gamma'_{xy} = u,_y + v,_x \tag{1.3}$$

For abbreviation, the denotation is used, e.g.

$$u,_x \triangleq \frac{\partial u}{\partial x} .$$

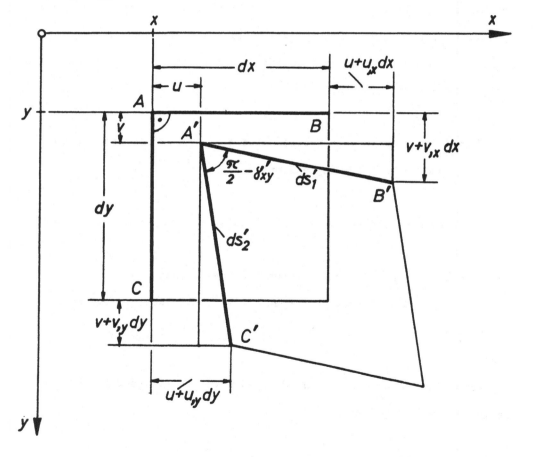

Fig. 2: In-plane deformations in the neutral surface

The strain-displacement relations by bending are derived from fig. 3:

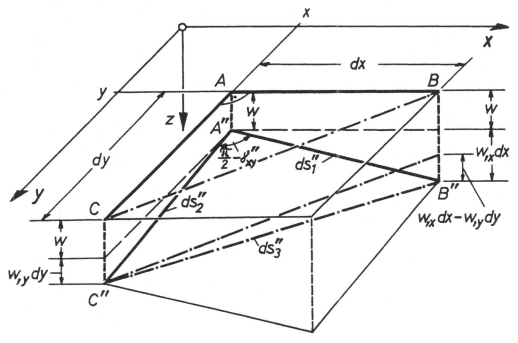

Fig. 3: In-plane deformations caused by bending

$$ds_1'' = dx \ (1 + w,_x^2)^{1/2}, \ \simeq dx \ (1 + \tfrac{1}{2} \ w,_x^2 + ---) \ ;$$

$$ds_2'' = dy \ (1 + w,_y^2)^{1/2}, \ \simeq dy \ (1 + \tfrac{1}{2} \ w,_y^2 + ---) \ ;$$

$$ds_3'' = dx^2 + dy^2 + (w,_x \cdot dx - w,_y \cdot dy)^2$$

$$\qquad = (ds_1'')^2 + (ds_2'')^2 - 2 \ ds_1'' \cdot ds_2'' \cdot \cos \ (\tfrac{\pi}{2} - \gamma_{xy}'')$$

$$\cos \ (\tfrac{\pi}{2} - \gamma_{xy}'') \ \simeq \ \gamma_{xy}$$

$$\varepsilon_x'' = \frac{1}{2} w_{,x}^2 , \tag{2.1}$$

$$\varepsilon_y'' = \frac{1}{2} w_{,y}^2 , \tag{2.2}$$

$$\gamma_{xy}'' = w_{,x} \cdot w_{,y} . \tag{2.3}$$

In the neutral plane, the following strain displacement relations are then obtained:

$$\varepsilon_x = u_{,x} + \frac{1}{2} w_{,x}^2 \tag{3.1}$$

$$\varepsilon_y = v_{,y} + \frac{1}{2} w_{,y}^2 \tag{3.2}$$

$$\gamma_{xy} = u_{,y} + v_{,x} + w_{,x} \cdot w_{,y} . \tag{3.3}$$

The compatibility conditions yield

$$\varepsilon_{x,yy} + \varepsilon_{y,xx} - \gamma_{xy,xy} = w_{,xy}^2 - w_{,xx} \cdot w_{,yy} \tag{4}$$

or after introducing the curvatures κ

$$\varepsilon_{x,yy} + \varepsilon_{y,xx} - \gamma_{xy,xy} = \kappa_{xy}^2 - \kappa_{xx} \cdot \kappa_{yy} \tag{4a}$$

In a distance z from the neutral plane, the strains by bending (fig. 4a/b) are obtained to

$$\varepsilon_x^b(z) = - z \cdot \kappa_{xx}; \quad \varepsilon_y^b(z) = - z \cdot \kappa_{yy}; \quad \gamma_{xy}^b(z) = - 2z \cdot \kappa_{xy} \tag{5}$$

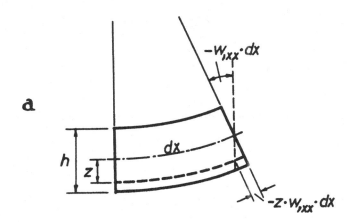

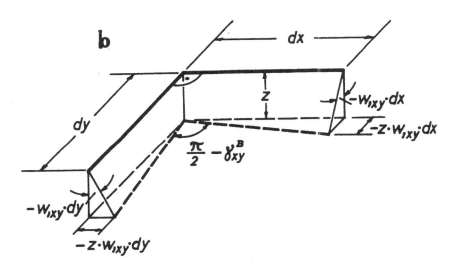

Fig. 4: Deformations depending on the coordinate z

and furthermore including the in-plane strains

$$\varepsilon_x(z) = u,_x + \frac{1}{2} w,_x^2 - z \cdot \kappa_{xx} \qquad (6.1)$$

$$\varepsilon_y(z) = v,_y + \frac{1}{2} w,_y^2 - z \cdot \kappa_{yy} \tag{6.2}$$

$$\gamma_{xy}(z) = u,_y + v,_x + w,_x \cdot w,_y - 2z \cdot \kappa_{xy} \tag{6.3}$$

Then the stress displacement relations in the neutral plane are formulated according to Hooke's Law of Elasticity:

$$\sigma_x = \frac{E}{1-\mu^2} [u,_x + \frac{1}{2} w,_x^2 + \mu(v,_y + \frac{1}{2} w,_y^2)], \tag{7.1}$$

$$\sigma_y = \frac{E}{1-\mu^2} [v,_y + \frac{1}{2} w,_y^2 + \mu(u,_x + \frac{1}{2} w,_x^2)], \tag{7.2}$$

$$\tau_{xy} = \frac{E}{2(1+\mu)}[u,_y + v,_x + w,_x \cdot w,_y] \tag{7.3}$$

and in a distance z from the neutral plane

$$\sigma_x(z) = \frac{E}{1-\mu^2} [u,_x + \frac{1}{2} w,_x^2 - z \cdot \kappa_{xx} +$$

$$+ \mu(v,_y + \frac{1}{2} w,_y^2 - z \cdot \kappa_{yy})], \tag{8.1}$$

$$\sigma_y(z) = \frac{E}{1-\mu^2} [v,_y + \frac{1}{2} w,_y^2 - z \cdot \kappa_{yy} +$$

$$+ \mu(u,_x + \frac{1}{2} w,_x^2 - z \cdot \kappa_{xx})], \tag{8.2}$$

$$\tau_{xy}(z) = \frac{E}{2(1+\mu)} [u,_y + v,_x + w,_x \cdot w,_y - 2z \cdot \kappa_{xy}] \tag{8.3}$$

with E the Young's Modulus and μ the Poisson Ratio.

The equilibrium conditions (fig. 5) yield

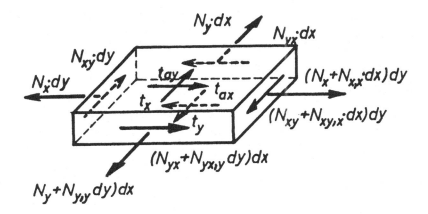

Fig. 5: In-plane forces and friction stresses

$$\Sigma X = 0: \quad N_{x,x} + N_{xy,y} + (t_{ax} - t_x) = 0 \tag{9.1}$$

$$\Sigma Y = 0: \quad N_{y,y} + N_{xy,x} + (t_{ay} - t_y) = 0 \tag{9.2}$$

with t_{ax}, t_{ay} the external loading in the xy-plane and t_x, t_y the friction stresses in the interface.

Introducing the functions

$$\phi_x = \int (t_{ax} - t_x)\, dx + C_1(y) \tag{10.1}$$

$$\rightarrow \quad t_{ax} - t_x = \phi_{x,x}$$

$$\Phi_y = \int (t_{ay} - t_y) \, dy + C_2(x) \tag{10.2}$$

$$\rightarrow \quad t_{ay} - t_y = \Phi_{y,y}$$

and a stress function F in such a way, that

$$N_x = F,_{yy} \cdot h - \Phi_x \tag{11.1}$$

$$N_y = F,_{xx} \cdot h - \Phi_y \tag{11.2}$$

$$N_{xy} = - F,_{xy} \cdot h \, , \tag{11.3}$$

the equilibrium conditions (9) are satisfied identically.

Integrating the stresses, eq.s (8), over the thickness h of the plate yields the internal forces, especially the bending and twisting moments.

$$M_x = - B (\kappa_{xx} + \mu \cdot \kappa_{yy}) \tag{12.1}$$

$$M_y = - B (\kappa_{yy} + \mu \cdot \kappa_{xx}) \tag{12.2}$$

$$M_{xy} = - B (1-\mu) \kappa_{xy} \tag{12.3}$$

where B denotes the bending stiffness of the plate

$$B = \frac{E \cdot h^3}{12(1-\mu^2)} \, .$$

Formulating the equilibrium conditions $\Sigma M = 0$, referred to the x-axis and the y-axis respectively, the relations are obtained

$$Q_x = - B \, (\nabla^2 w),_x - (t_{ax} + t_x) \, \frac{h}{2} \tag{13.1}$$

$$Q_y = - B \, (\nabla^2 w),_y - (t_{ay} + t_y) \, \frac{h}{2} \tag{13.2}$$

with the Laplace operator ∇^2:

$$\nabla^2 w = w,_{xx} + w,_{yy} \, .$$

Neglecting terms, which are getting small in higher order, the equilibrium condition in direction of the z-axis yields (fig. 6) with reference to the deformed element in the xz-plane (p_a and p will be considered in eq. (17)

$$Z = N_x \cdot w,_{xx} \cdot dxdy + Q_{x,x} \cdot dxdy \tag{14}$$

with the approximation $\sin\varphi_x \simeq \varphi_x$, $\cos\varphi_x \simeq 1$, and in the yz-plane

$$Z = N_y \cdot w,_{yy} \cdot dxdy + Q_{y,y} \cdot dxdy \, . \tag{15}$$

According to fig. 7 will be obtained

$$Z = 2 \cdot N_{xy} \cdot w,_{xy} \cdot dxdy \tag{16}$$

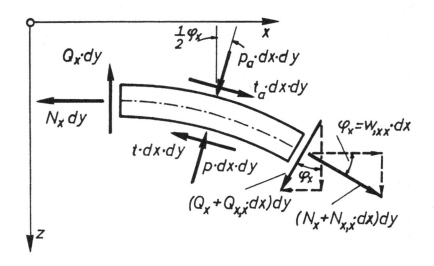

Fig. 6: Equilibrium condition in direction of the z-axis

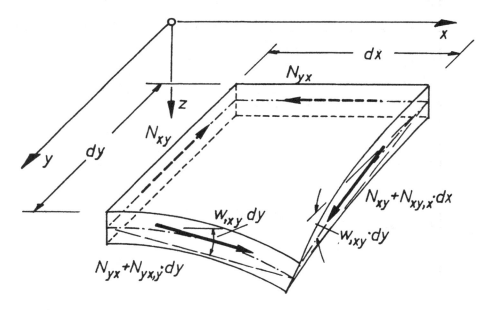

Fig. 7: Influence of N_{xy} on the equilibrium condition $\Sigma Z = 0$

Regarding eq.s (14), (15) and (16), the condition $\Sigma Z = 0$ then runs

$$Q_{x,x} + Q_{y,y} + N_x \cdot w_{,xx} + N_y \cdot w_{,yy}$$

$$+ 2N_{xy} \cdot w_{,xy} + (p_a - p) = 0 . \qquad (17)$$

Introducing the stress function, eq.s (11) and the relations (13), the equation (17) runs

$$\frac{B}{h} \nabla^2 \nabla^2 w = L(w,F) - \frac{1}{2}(t_{ax} + t_x)_{,x} - \frac{1}{2}(t_{ay} + t_y)_{,y}$$

$$- \frac{1}{h}(\phi_x \cdot \kappa_{xx} + \phi_y \cdot \kappa_{yy}) + \frac{1}{h}(p_a - p) , \qquad (18)$$

where $L(w,F)$ denotes a differential operator

$$L(w,F) = w_{,xx} \cdot F_{,yy} + w_{,yy} \cdot F_{,xx} - 2w_{,xy} \cdot F_{,xy} .$$

Regarding Hooke's Law of Elasticity, the compatibility condition (4a) runs

$$\sigma_{x,yy} - \mu \cdot \sigma_{y,yy} + \sigma_{y,xx} - \mu \cdot \sigma_{x,xx} - 2(1+\mu) \tau_{xy,xy}$$

$$= E(\kappa_{xy}^2 - \kappa_{xx} \cdot \kappa_{yy}) \qquad (19)$$

and furthermore, introducing the stress function

$$\frac{1}{E} \nabla^2\nabla^2 F = -\frac{1}{2} L(w,w) + \frac{1}{Eh} (\Phi_{x,yy} + \Phi_{y,xx}) - $$

$$- \frac{\mu}{Eh} (\Phi_{x,xx} + \Phi_{y,yy}) \qquad\qquad (20)$$

Equations (18) and (20) are coupled non-linear differential equations describing a geometric non-linear plate theory regarding large deflections, contact to a yielding subgrade and including friction stresses in the contact area.

These equations can also be derived in case of orthotropic plates and plates with variable thickness $h(x,y)$.

Although the derived constitutive equations must still be considered as an approach, the real conditions are described much closer to reality by this extended plate theory than by the classical Kirchhoff's theory

$$B \nabla^2\nabla^2 w = (p_a - p) , \qquad\qquad (21)$$

where the in-plane stress state is regarded to be independent of the bending stress state and vice versa.

For central-symmetrical plates it is useful to introduce polar coordinates, i.e. the only variable r. Assuming t_{ar} to be zero, the differential equations are running

$$\frac{B}{h} \nabla^2 \nabla^2 w = L\ (w,F) - \frac{1}{h}\ \Phi \cdot w,_{rr} - \frac{1}{2}\ \Phi,_{rr} + \frac{1}{h}\ (p_a - p) \qquad (22)$$

$$\frac{1}{E} \nabla^2 \nabla^2 F = -\frac{1}{2}\ L\ (w,w) + \frac{1}{Eh}\ (\frac{1}{r}\ \Phi,_r - \mu\Phi,_{rr}) \qquad (23)$$

with the Laplace operator

$$\nabla^2 w = w,_{rr} + \frac{1}{r}\ w,_r\ ,$$

the differential operator

$$L\ (w,F) = w,_{rr} \cdot \frac{1}{r}\ F,_r + \frac{1}{r}\ w,_r \cdot F,_{rr}\ ,$$

and the function of the friction stresses

$$\Phi = -\int_o^r t(\bar{r}) \cdot d\bar{r} + C\ .$$

Introducing the stress resultants of the in-plane stress state, eq.s (22) and (23) run

$$B \nabla^2 \nabla^2 w = N_r \cdot w,_{rr} + N_\varphi \cdot \frac{1}{r}\ w,_r - \frac{h}{2}\ \Phi,_{rr} + (p_a - p) \qquad (24)$$

$$\frac{1}{E} \nabla^2\ (N_r + N_\varphi + \Phi) = -\frac{1}{r}\ w,_r \cdot w,_{rr} \cdot h + \frac{1}{E}\ (\frac{1}{r}\ \Phi,_r - \mu\Phi,_{rr}) \qquad (25)$$

The differential equation of the 4th order (24) is split up into two differential equations of the 2nd order

$$\nabla^2 w = -\frac{M}{B} , \tag{26.1}$$

$$\nabla^2 M = - N_r \cdot w,_{rr} - N_\varphi \cdot \frac{1}{r} \cdot w,_r + \frac{h}{2} \Phi,_{rr} - (p_a - p) , \tag{26.2}$$

where M denotes the invariant sum of the bending moments.

For mathematical solution, the stress state and the state of displacements of the yielding subgrade, e.g. the elastic halfspace, must be described, and the contact conditions must be introduced. Generally valid solutions of this theory have not yet been found. Therefore the hybrid technique will be introduced to take characteristic data by experiment as input data for the differential equations of the plate only after transforming them into finite formulas. Then the final solution including the contact stresses and the contact areas can be determined by simple matrix calculus.

The transformation of the differential equations yields

a) in Cartesian coordinates

$$\underset{\sim}{D} \cdot \underset{\sim}{m} = \lceil \underset{\sim}{n}_x \rfloor \; \underset{\sim}{w}'' - \lceil \underset{\sim}{n}_y \rfloor \cdot \underset{\sim}{w}^{\cdot\cdot} - 2\lceil \underset{\sim}{n}_{xy} \rfloor \cdot \underset{\sim}{w}'^{\cdot} +$$

$$+ \frac{h}{2} (\Phi_x'' + \Phi_y^{\cdot\cdot}) - \underset{\sim}{p}_a + \underset{\sim}{p} \tag{27.1}$$

$$\underset{\sim}{D}\,(\underset{\sim}{n} + \underset{\sim}{\phi}) = - E \cdot h \left(\lceil \underset{\sim}{w}'' \rfloor \cdot \underset{\sim}{w}^{\cdot\cdot} - \lceil \underset{\sim}{w}^{\cdot\cdot} \rfloor \cdot \underset{\sim}{w}^{\cdot\cdot} \right) +$$

$$+ (\underset{\sim x}{\phi}^{\cdot\cdot} + \underset{\sim y}{\phi}'') - \mu(\underset{x}{\phi}'' + \underset{y}{\phi}^{\cdot\cdot}) \tag{27.2}$$

with $\underset{\sim}{m}$ the vector $\{M_{ij}\}$,

$\underset{\sim}{n}$ the vector $\{(N_x + N_y)_{ij}\}$,

$\underset{\sim}{w}''$ the vector of the second partial derivative $\underset{\sim}{w}'' = \{(w,_{xx})_{ij}\}$ and
$$\underset{\sim}{w}^{\cdot\cdot} = \{(w,_{yy})_{ij}\}$$

respectively. The symbol $\lceil --- \rfloor$ denotes a diagonal matrix, so that e.g.

$\lceil \underset{\sim x}{n} \rfloor \cdot \underset{\sim}{e} = \underset{\sim x}{n}$, where $\underset{\sim}{e}$ denotes the unit vector.

b) in polar coordinates

$$\underset{\sim}{D} \cdot \underset{\sim}{m} = - \lceil \underset{\sim r}{n} \rfloor \underset{\sim}{w}'' - \lceil \underset{\sim \varphi}{n} \rfloor \cdot \underset{\sim}{w}' + \frac{h}{2} \underset{\sim}{\phi}'' - \underset{\sim a}{p} + \underset{\sim}{p} \tag{28.1}$$

$$\underset{\sim}{D}\,(\underset{\sim r}{n} + \underset{\sim \varphi}{n} + \underset{\sim}{\phi}) = - Eh \lceil \underset{\sim}{w}' \rfloor \underset{\sim}{w}'' + \underset{\sim}{\phi}' - \mu \cdot \underset{\sim}{\phi}'' \tag{28.2}$$

with $\underset{\sim}{w}'' = \{(w,_{rr})_i\}$, $\underset{\sim}{w}' = \{(\frac{1}{r} w,_r)_i\}$ and the derivatives of ϕ respectively.

To analyze the stress state as well as the contact stresses in plates on yielding subgrades for arbitrary boundary conditions and/or arbitrary conditions of support, as demonstrated in fig. 8 by some examples, photo-elastic and Moiré methods or a combination of both methods can be used

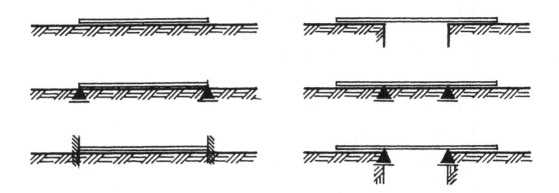

Fig. 8: Examples of plate support

as well as strain gauge measurements. Generally, no restrictions exist with respect to the geometry of the plates. According to the previously made assumptions, linear-elastic, homogeneous and isotropic response of material is supposed.

3. Photoelasticity

Based on the well-known experimental method of photoelasticity (see e.g. Frocht[3], Wolf[4], Laermann[2]), the basic relations related to a one-dimensional stress state will be considered at first, which is described by the bending moment M and the internal force N.

Considering a single-layer model (fig. 9) with the mirrored surface turned to the subgrade, the birefringence effect observed by a reflection

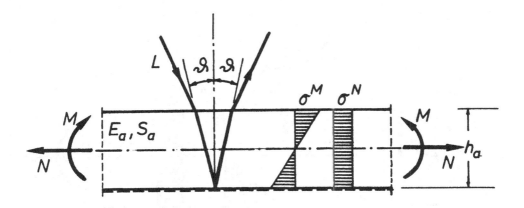

Fig. 9: Single-layer model; stresses and light path

polariscope is influenced by the internal force N only. Minor influences of the stress gradient and boundary effects on the radiation are neglected. The observed isochromatic fringe order δ runs

$$\delta_N = \frac{2}{S_a} N \ , \tag{29}$$

where S_a denotes the stress-optical response of the model material, assumed to depend on the wave length of the light only.

Considering a two-layer model (fig. 10) with a reflecting interface between the two layers of different mechanical behavior as well as optical response, the relation between the order of birefringence and the stress runs

$$\delta_{M,N} = 2 \int_o^{h_o} \sigma_o(\bar{z}) \ \frac{d\bar{z}}{S_b} \tag{30}$$

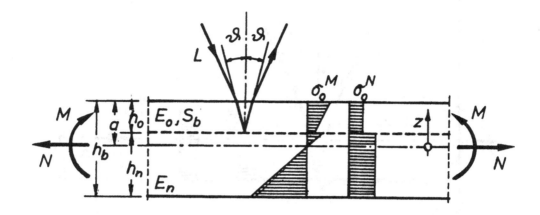

Fig. 10: Two-layer model; stresses and light path

where now the bending stress as well as the normal stress caused by N are
of influence on the retardation in the layer of photoelastic material.
With

$$\sigma_o(\bar{z}) = \sigma_o{}^M \left(1 - \frac{h_o}{a} + \frac{\bar{z}}{a}\right) + \sigma_o{}^N \ , \tag{31}$$

where a is the distance of the neutral plane from the upper surface,

$$a = \frac{h_o{}^2 + E_n/E_o \cdot (h_b{}^2 - h_o{}^2}{2[h_o + E_n/E_o \ (h_b - h_o)]} \ , \tag{32}$$

relation (30) holds

$$\delta_{M,N} = \frac{2}{S_b} \ h_o \ [\ \sigma_o{}^M \left(1 - \frac{h_o}{2a}\right) + \sigma_o{}^N \ ] \ . \tag{33}$$

Between the internal forces

$$N = \int_{-(h_b - a)}^{a} \sigma(z) \cdot dz \;, \quad M = \int_{-(h_b - a)}^{a} \sigma(z) \cdot z \cdot dz \qquad (34)$$

and the boundary values of the stresses, the relations are valid:

$$\sigma_o^{\;N} = N \; / \; [h_o + \frac{E_n}{E_o} (h_b - h_o)] \qquad (35.1)$$

$$\sigma_o^{\;M} = M \cdot 3a \; / \; \{a^3 - (a - h_o)^3 + \frac{E_n}{E_o} [(a - h_o)^3 + (h_b - a)^3]\} \;. \qquad (35.2)$$

Then the relation between the order of birefringence and the internal forces runs

$$\delta_{M,N} = \frac{2h_o}{S_b} \{ \; N \; / \; [h_o + \frac{E_n}{E_o} (h_b - h_o)] \quad +$$

$$+ \; M \cdot [3a(1 - h_o/2a) \;] \; / \; \{a^3 - (a - h_o)^3 + \qquad (36)$$

$$+ \frac{E_n}{E_o} [(a - h_o)^3 + (h_b - a)^3]\} \; \}$$

Introducing the calibration factors

$$K_N = \frac{S_b}{4h_o} \; [h_o + \frac{E_n}{E_o} (h_b - h_o)] \qquad (37.1)$$

$$K_M = \frac{S_b}{12h_o} \cdot \{a^3 - (a - h_o)^3 + \frac{E_n}{E_o} [(a - h_o)^3 +$$

$$+ \; (h_b - a)^3]\}/[3a(1 - h_o/2a)] \;, \qquad (37.2)$$

equation (36) runs

$$\delta_{M,N} = \frac{N}{2K_N} + \frac{M}{2K_M} \tag{38}$$

K_N and K_M are to be determined in two separate calibration tests.

Condition: Equal tensile and bending stiffness of the single-layer and the two-layer model.

If this condition is satisfied, eq. (29) will be introduced into eq. (38); then follows

$$M = 2K_M \left(\delta_{M,N} - \frac{S_a}{4K_N} \delta_N\right) = 2K_M \cdot \delta_M \tag{39}$$

with

$$\delta_M = \delta_{M,N} - \frac{S_a}{4K_N} \cdot \delta_N \; . \tag{40}$$

Now a two-dimensional stress state will be considered, where an in-plane stress state described by maxN, the main shear stress resultant, is superimposed by the bending stress state described by $maxM_D$, the main twisting moment.

Photoelastic experiment of a single-layer model yields

$$\text{maxN} = S_a \cdot \delta_N . \tag{41}$$

Under the already previously mentioned condition of equal in-plane
and bending stiffness of both the models, and under supposition, that the
principal axes of the bending and of the in-plane stress state coincide
from the two-layer model the experimental data are obtained

$$\delta_{M,N} = \text{maxN}/K_N + \text{maxM}_D/K_M . \tag{42}$$

It must be mentioned here, that the relation (42) is valid under the assumption, that the neutral planes of the normal stresses and the shear
stresses coincide.

From eq.s (41) and (42) follows

$$\text{maxM}_D = K_M \left(\delta_{M,N} - \frac{S_a}{K_N} \delta_N \right) = K_M \cdot \delta_M . \tag{43}$$

Generally, in case of superimposed in-plane and bending stress state
the principal stress directions are not constant over the thickness of
the model and the photoelastic layer, but a rotation will be observed
along the direction of light propagation. Several authors have dealt with
that problem. Drucker and Mindlin[5] have given a solution related to a
special case. They showed that, when the incoming light vector is linearly polarized and oriented parallel to a principal stress direction at the

model surface facing the light source, the light vector remains linearly polarized and follows the orientation of the principal stress direction throughout the thickness of the model. But this is valid only, if the ratio of rotation to the phase retardation is small. Kuske[6] derived a graphical method by projecting the Poincaré sphere on to the equatorial plane (j-circle). Aben[7] and other references have presented a numerical method based on matrix calculus. He has introduced the "characteristic directions".

These polarization directions, in general, do not coincide with the axes of principal stress, however their experimental determination gives additional information about the stress state in the model.

For the single-layer model as well as for the two-layer model, the principal stress directions can be described by a function of the dimensionless coordinate $\xi = z/a$ in direction of light propagation:

$$\tan 2\psi(\xi) = \frac{\tan 2\psi^M \cdot \xi + Q \cdot \tan 2\psi^N}{Q + \xi} \, , \tag{44}$$

where ψ^M denotes the principal direction of the bending stress state, ψ^N that of the in-plane stress state. For the single-layer model, Q holds

$$Q_a = \frac{maxN \cdot \cos 2\psi^N}{maxM_D \cdot \cos 2\psi^M} \Big/ \frac{6}{h_a} \, . \tag{45}$$

and for the two-layer model with respect to eq.s (35)

$$Q_b = \frac{maxN \cdot \cos 2\psi^N}{maxM_D \cdot \cos 2\psi^M} \left/ \frac{3a[h_o + \frac{E_n}{E_o}(h_b - h_o)]}{a^3 - (a-h_o)^3 + \frac{E_n}{E_o}[(a-h_o)^3 + (h_b - a)^3]} \right. \tag{46}$$

As both the models are observed in a reflection polariscope, the angle α_1

(fig. 11) as the characteristic direction at the upper plate surface is

obtained.

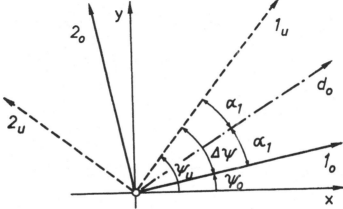

Fig. 11: Principal axes, characteristic direction and angle of
rotation

Then the difference $\Delta\psi$ of the principal directions between the upper and

the lower surface equals $2\alpha_1$:

$$\Delta\psi = \psi_u - \psi_o = 2\alpha_1 . \tag{47}$$

From eq. (44) follows

$$\Delta\psi = \frac{1}{2} \text{ arc tan } \frac{2Q \cdot a(\tan 2\psi^N - \tan 2\psi^M)}{(Q^2 - a^2) + Q^2 \cdot \tan^2 2\psi^N - a^2 \cdot \tan^2 2\psi^M} \tag{48}$$

The data $\Delta\psi_a$ and $\Delta\psi_b$ are obtained from both the experiments using linearly polarized monochromatic light, as it can be shown, that the characteristic directions depend on the wave length of the used light.

The relation between the observed retardation δ and the stresses can be expressed for the single-layer model as follows:

$$\delta_a = 48 \cdot \frac{1}{S_a} \cdot \frac{1}{h_a^3} \cdot \max M_D \cdot \cos 2\psi^M \left\{ \int_{-1}^{+1} \frac{\xi}{\cos 2\psi(\xi)} \, d\xi \right.$$
$$\left. + Q_a \cdot \int_{-1}^{+1} \frac{d\xi}{\cos 2\psi(\xi)} \right\} \tag{49}$$

with $\xi = \frac{z}{a}$, $a = \frac{h_a}{2}$,

and for the two-layer model

$$\delta_b = 12 \cdot \frac{1}{S_b} \cdot \frac{\max M_D \cdot \cos 2\psi^M}{a^3 - (a-h_o)^3 + \frac{E_n}{E_o}[(a-h_o)^3 + (h_b-a)^3]}$$
$$\left\{ \int_{\frac{a-h_o}{a}}^{1} \frac{\xi}{\cos 2\psi(\xi)} \, d\xi + Q_b \int_{\frac{a-h_o}{a}}^{1} \frac{d\xi}{\cos 2\psi(\xi)} \right\} \tag{50}$$

with a according to eq. (32).

From eq. (44) follows

$$\cos 2\psi(\xi) = \frac{Q + \xi}{[a \cdot \xi^2 + b\xi + c]^{1/2}} , \tag{51}$$

where a, b, c are coefficients depending on Q, ψ^M and ψ^N. The solution of the integrals eq.s (49) and (50) regarding eq. (51) yields the relation between the observed retardation and the in-plane as well as the

bending stress state.

Four experimental data: $\Delta\psi_a$, $\Delta\psi_b$, δ_a and δ_b are now available to de-

termine the four unknowns $\max M_D$, $\max N$, ψ^M and ψ^N. For further evaluation

of these experimental data, the well-known Frocht's[3] shear stress differ-

ence method is introduced to determine the in-plane stress state. As the

stress state depends on the friction stresses in the interface as well

(fig. 5), the relations are valid

$$N_x = - \int_0^x N_{xy,y} \cdot d\xi + \int_0^x t_x \cdot d\xi + C(y) \tag{52.1}$$

$$N_y = - \int_0^y N_{xy,x} \cdot d\eta + \int_0^y t_y \cdot d\eta + C(x) \tag{52.2}$$

After integration and introducing the functions Φ_x and Φ_y, eq.s (52) run

$$(\bar{N}_x)_i = (N_x)_i + (\Phi_x)_i = - \sum_{v=1}^{i} (N_{xy,y})_v \cdot \Delta x + (N_x)_o \tag{53.1}$$

for $i \in [1/n]$ and all $y = y_j = j \cdot \Delta y$, $j \in [1/m]$,

$$(\bar{N}_y)_j = (N_y)_j + (\Phi_y)_j = - \sum_{v=1}^{j} (N_{xy,x})_v \cdot \Delta y + (N_y)_o \tag{53.2}$$

for $j \in [1/m]$ and all $x = x_i = i \cdot \Delta x$, $i \in [1/n]$.

In these equations, $(N_{xy,y})_i$ and $(N_{xy,x})_j$ denote the derivations $\frac{\partial N_{xy}}{\partial y}$ and $\frac{\partial N_{xy}}{\partial x}$ respectively with

$$N_{xy} = maxN \cdot \sin 2\psi^N \qquad (54)$$

It must be pointed out that $\bar{N}_{xy} = N_{xy}$.

If no contact problem is to consider, the stress resultants of the in-plane stresses run

$$(N_x)_i = - \sum_{\nu=1}^{i} (N_{xy,y})_\nu \cdot \Delta x + (N_x)_o \qquad (55.1)$$

$$(N_y)_i = (N_x)_i \mp 2\, maxN_i \cdot \cos 2\psi_i^N \qquad (55.2)$$

To analyze the bending stress state in a Cartesian coordinate system, the method of Haberland[8] is applied.

The following relations between the bending moments and the twisting moment are valid

$$(M_x - \mu M_y),_y = (1 + \mu)\, M_{xy,x} \;, \qquad (56.1)$$

$$(M_y - \mu M_x),_x = (1 + \mu)\, M_{xy,y} \;. \qquad (56.2)$$

Introducing the "equivalent moments"

$$m_x = M_x - \mu \cdot M_y \; , \tag{57.1}$$

$$m_y = M_y - \mu \cdot M_x \tag{57.2}$$

and integrating eq.s (56), these run

$$m_x = (1 + \mu) \int M_{xy,x} \; dy + C_1(x) \; , \tag{58.1}$$

$$m_y = (1 + \mu) \int M_{xy,y} \; dx + C_2(y) \tag{58.2}$$

or, transformed into summation equations considering the boundary conditions

$$(m_x)_j = (1 + \mu) \cdot \sum_{\nu=1}^{j} (M_{xy,x})_\nu \cdot \Delta y + (m_x)_o \tag{59.1}$$

for $j \in [1/m]$ and all $x = x_i = i \cdot \Delta x$, $i \in [1/n]$

$$(m_y)_i = (1 + \mu) \cdot \sum_{\nu=1}^{i} (M_{xy,y})_\nu \cdot \Delta x + (m_y)_o \tag{59.2}$$

for $i \in [1/n]$ and all $y = y_i = j \cdot \Delta y$, $j \in [1/m]$.

With the experimental data $\text{max}M_D$ and ψ^M, the twisting moment related to a Cartesian coordinate system runs

$$(M_{xy})_{ij} = \text{max}M_{D_{ij}} \; \sin 2\psi^M_{ij} \; . \tag{60}$$

Having determined the equivalent moments according to eq.s (59) considering eq. (60), the bending moments run

$$(M_x)_{ij} = \frac{1}{1-\mu} [\, (m_y)_{ij} \pm 2\, maxM_{D_{ij}} \cdot \cos 2\psi^M_{ij}\,] , \qquad (61.1)$$

$$(M_y)_{ij} = \frac{1}{1-\mu} [\, (m_y)_{ij} \pm \mu \cdot 2\, maxM_{D_{ij}} \cdot \cos 2\psi^M_{ij}\,] , \qquad (61.2)$$

and furthermore the sum of the bending moments

$$M_{ij} = \frac{2}{1-\mu^2} [\, (m_y)_{ij} \pm (1 + \mu)\, maxM_{D_{ij}} \cdot \cos 2\psi^M_{ij}\,] . \qquad (62)$$

Regarding the relations between the bending and the twisting moments respectively and the deflection w, the second derivatives of w can be expressed as follows

$$w,_{xx} = -\frac{m_x}{B(1-\mu^2)} , \qquad (63.1)$$

$$w,_{yy} = -\frac{m_y}{B(1-\mu^2)} , \qquad (63.2)$$

$$w,_{xy} = -\frac{M_{xy}}{B(1-\mu)} . \qquad (63.3)$$

In order to obtain higher accuracy in the final results by smoothing and balancing the experimental data, these may be evaluated by means of spline functions.

Eqs. (53), (54), (62) and (63) yield the elements of the vectors and matrices respectively $\underset{\sim}{m}$, $\bar{\underset{\sim}{n}}$, $\bar{\underset{\sim}{n}}_x$, $\bar{\underset{\sim}{n}}_y$, $\underset{\sim}{n}_{xy} = \bar{\underset{\sim}{n}}_{xy}$, $\underset{\sim}{w}''$, $\underset{\sim}{w}^{\cdot\cdot}$ and $\underset{\sim}{w}'^{\cdot}$, as have been introduced in eq.s (27).

The numerical evaluation of eq. (27.2) yields the sum $t_{ij} = t_{xij} + t_{yij}$ of the components of the friction stresses in direction of the coordinate axes x and y, regarding the meaning of $\bar{n}$:

$$\underset{\sim}{t} = \underset{\sim}{T}^{-1} [\underset{\sim}{D} \cdot \bar{\underset{\sim}{n}} + (E \cdot h) (\lceil \underset{\sim}{w}'' \rfloor \cdot \underset{\sim}{w}^{\cdot\cdot} - \lceil \underset{\sim}{w}'^{\cdot} \rfloor \cdot \underset{\sim}{w}'^{\cdot})] . \qquad (64)$$

The difference $N_x - N_y$ with reference to eq.s (52) holds with $t_y = t - t_x$

$$N_x - N_y = - \int_x N_{xy,y} \, dx + \int_y N_{xy,y} \, dy +$$
$$+ \int_x t_x \cdot dx + \int_y t_x \cdot dy - \int_y t \cdot dy + (N_x - N_y)_o . \qquad (65)$$

With respect to eq. (54), the difference $N_x - N_y$ can be expressed by the experimental data

$$N_x - N_y = \frac{1}{2} N_{xy} / \tan 2\psi^N = \frac{1}{2} maxN \cdot \cos 2\psi^N . \qquad (66)$$

Then eq. (65) yields, considering the boundary conditions,

$$\int_x t_x \, dx + \int_y t_x \, dy = (\frac{1}{2} \max N \cdot \cos 2\psi^N) - (\frac{1}{2} \max N_o \cdot \cos 2\psi^N_o) +$$

$$+ \int_x N_{xy,y} \, dx - \int_y N_{xy,x} \, dy + \int_y t \cdot dy \tag{67}$$

Integration of eq. (67) along y = const leads to the recurrent formula to determine the friction stress component t_x

$$t_{xi} = K_i \frac{1}{\Delta x} + \sum_{\nu=1}^{i} (N_{xy,y})_\nu - \sum_{\nu=1}^{i-1} t_{x\nu} \tag{68}$$

for i ∈[1/n] and all y = y_j, j ∈[1/m]

with

$$K_i = \frac{1}{2} [\max N_i \cdot \cos 2\psi^N_i - \max N_o \cdot \cos 2\psi^N_o] .$$

Thus, the components t_x and t_y in discrete points ij may be determined, and furthermore N_x as well as N_y. Then from eq. (27.1), the solution vector $\underset{\sim}{p}$ follows

$$\underset{\sim}{p} = \underset{\sim}{p_a} + D \cdot \underset{\sim}{m} - \lceil n_x \rfloor \, \underset{\sim}{w}'' + \lceil n_y \rfloor \, \underset{\sim}{w}^{\cdot\cdot} + 2\lceil n_{xy} \rfloor \cdot \underset{\sim}{w}'^{\cdot} +$$

$$+ \frac{h}{2} (\Phi_x'' + \Phi_y^{\cdot\cdot}) . \tag{69}$$

For the evaluation of eq.s (64), (68) and (69), a computer program is necessary.

In the special case of a central-symmetrically loaded circular plate, the principal stress directions of both the stress states (in-plane and bending stress state) coincide. Then the relations (41), (42) and (43) are valid. Introducing polar coordinates (r, φ), the evaluation of the data δ_N taken from the single-layer model yields the in-plane stress resultants N_r and N_φ as the resultants of the principal stresses. The angle ψ^N and ψ^M of the principal axes are zero with reference to the angle φ of the polar coordinates. Then the difference of principal stress resultants runs

$$N_r - N_\varphi = \frac{S_a}{2} \, \delta_N \; . \tag{70}$$

The equilibrium condition in direction of r holds

$$dN_r = -\frac{1}{r} (N_r - N_\varphi) \, dr + t \; . \tag{71}$$

With eq. (70), integration of eq. (71) yields

$$N_r(r) = -\frac{S_a}{2} \int_r \delta_N(\bar{r}) \, \frac{d\bar{r}}{\bar{r}} + \int_r t(\bar{r}) \, d\bar{r} + C_1 \; , \tag{72.1}$$

$$N_\varphi(r) = -\frac{S_a}{2} \{ \, \delta_N(r) + \int_r \delta_N(\bar{r}) \, \frac{d\bar{r}}{\bar{r}} \} + \int_r t(\bar{r}) \, d\bar{r} + C_2 \; . \tag{72.2}$$

The friction stresses t are still unknown, therefore from experiment it can be obtained only

$$\bar{N}_r(r) = N_r(r) + \Phi(r) = - \frac{S_a}{2} \int_r \delta_N(\bar{r}) \frac{d\bar{r}}{r} + C_1 \ , \tag{73.1}$$

$$\bar{N}_\varphi(r) = N_\varphi(r) + \Phi(r) = - \frac{S_a}{2} \{ \ \delta_N(r) + \int_r \delta_N(\bar{r}) \frac{d\bar{r}}{r} \ \} + C_2 \ . \tag{73.2}$$

For numerical evaluation, eq.s (73) are transformed into a finite form:

$$(\bar{N}_r)_i = - \frac{S_a}{2} \sum_{\nu=n}^{i} \frac{1}{\nu} (\delta_N)_\nu \cdot \Delta + (\bar{N}_r)_n \ , \tag{74.1}$$

$$(\bar{N}_\varphi)_i = - \frac{S_a}{2} \{ \ (\delta_N)_i + \sum_{\nu=n}^{i} \frac{1}{\nu} (\delta_N)_\nu \cdot \Delta \ \} + (\bar{N}_\varphi)_n \ , \tag{74.2}$$

with $\Delta = \Delta r/r = 1/n$.

In matrix form, eq.s (74) run

$$\bar{\underset{\sim}{n}}_r = \{ \ (N_r)_i \ \} = - \frac{S_a}{2} \cdot \underset{\sim}{R}_N \cdot \underset{\sim}{\delta}_N \ , \tag{75.1}$$

$$\bar{\underset{\sim}{n}}_\varphi = \{ \ (N_\varphi)_i \ \} = - \frac{S_a}{2} \ (\underset{\sim}{R}_N + \underset{\sim}{I}) \ \underset{\sim}{\delta}_N \ . \tag{75.2}$$

$\underset{\sim}{I}$ denotes the unit matrix; the boundary conditions are included in the matrix R_N.

Evaluation of the bending stress state according to Laermann[9] yields the equivalent moments in polar coordinates

$$(m_\varphi)_i = 2(1 + \mu)K_M \sum_{\nu=n}^{i} \frac{1}{\nu} (\delta_M)_\nu \cdot \Delta + (m_\varphi)_n \tag{76.1}$$

$$(m_r)_i = 2(1 + \mu)K_M [\ (\delta_M)_i + \sum_{\nu=n}^{i} \frac{1}{\nu} (\delta_M)_\nu \cdot \Delta \ ] + (m_r)_n \tag{76.2}$$

where δ_M follows from eq. (43).

Introducing vectors and matrices, eq.s (76) are running in the following form:

$$\underset{\sim}{m}_\varphi = \{ (m_\varphi)_i \} = 2(1 + \mu) \cdot K_M \cdot \underset{\sim}{R}_M \cdot \underset{\sim}{\delta}_M \, , \tag{77.1}$$

$$\underset{\sim}{m}_r = \{ (m_r)_i \} = 2(1 + \mu) \cdot K_M (\underset{\sim}{R}_M + \underset{\sim}{I}) \underset{\sim}{\delta}_M \, . \tag{77.2}$$

The derivatives of the deflection surface w are expressed by the equivalent moments

$$w,_{rr} = - \frac{m_r}{B(1-\mu^2)} \; ; \quad \frac{1}{r} \cdot w,_r = - \frac{m_\varphi}{B(1-\mu^2)} \quad , \tag{78}$$

and in the matrix form with the vectors $\underset{\sim}{w}''$ and $\underset{\sim}{w}'$

$$\underset{\sim}{w}'' = - \frac{2 \, K_M}{B(1-\mu)} (\underset{\sim}{R}_M + \underset{\sim}{I}) \underset{\sim}{\delta}_M \, , \tag{79.1}$$

$$\underset{\sim}{w}' = - \frac{2 \, K_M}{B(1-\mu)} \cdot \underset{\sim}{R}_M \cdot \underset{\sim}{\delta}_M \, . \tag{79.2}$$

Then the sum of the bending moments runs in the matrix form

$$\underset{\sim}{m} = - \frac{2 \, K_M}{1-\mu} (2 \, \underset{\sim}{R}_M + \underset{\sim}{I}) \underset{\sim}{\delta}_M \, . \tag{80}$$

Note: The differences in the matrices $\underset{\sim}{R}_N$ and $\underset{\sim}{R}_M$ result from different boundary conditions.

Regarding eq.s (73) and (75) respectively as well as eq.s (79), the finite difference equation (28.2) holds

$$\underset{\sim}{t} = \underset{\sim}{T}^{-1} \left[-\frac{S_a}{2} \cdot \underset{\sim}{D} \left(2 \, \underset{\sim}{R}_N + \underset{\sim}{I} \right) \delta_N + \right.$$
$$\left. + E \cdot h \cdot \left(\frac{2 \, K_M}{B(1-\mu)}\right)^2 \lceil \, \underset{\sim}{R}_M \cdot \delta_M \, \rfloor \left(\underset{\sim}{R}_M + \underset{\sim}{I} \right) \delta_M \, \right] \tag{81}$$

with the non-singular square matrix

$$\underset{\sim}{T} = r_a \cdot \underset{\sim}{D} \cdot \underset{\sim}{A} + \frac{1}{r_a} \left(\underset{\sim}{A}_1 - \mu \cdot \underset{\sim}{A}_2 \right) . \tag{82}$$

With the results $\underset{\sim}{t} = \{(t_r)_i\}$, the in-plane stresses and their stress resultants respectively can be calculated. Then from eq. (28.1) follows the solution vector $\underset{\sim}{p}$ of the contact stresses normal to the contact interface

$$\underset{\sim}{p} = \underset{\sim}{p}_a + \underset{\sim}{D} \cdot \underset{\sim}{m} + \lceil \underset{\sim}{n}_r \rfloor \cdot \underset{\sim}{w}'' + \lceil \underset{\sim}{n}_\varphi \rfloor \cdot \underset{\sim}{w}' - \frac{h}{2} \cdot \underset{\sim}{\phi}'' . \tag{83}$$

With the experimental data δ_N and δ_M, eq. (83) runs

$$\underset{\sim}{p} = \underset{\sim}{p}_a - \frac{2 \, K_M}{1-\mu} \left\{ \left(\underset{\sim}{D} + \frac{r_a}{B} \lceil \, \underset{\sim}{A} \cdot \underset{\sim}{t} \, \rfloor \right) \left(2 \, \underset{\sim}{R}_M + \underset{\sim}{I} \right) - \right.$$
$$\left. - \frac{S_a}{2B} \left[\lceil \underset{\sim}{R}_N \cdot \delta_N \rfloor \left(\underset{\sim}{R}_M + \underset{\sim}{I} \right) + \lceil \left(\underset{\sim}{R}_N + \underset{\sim}{I} \right) \cdot \delta_N \rfloor \underset{\sim}{R}_M \right] \right\} \delta_M +$$
$$+ \frac{h}{2r_a} \underset{\sim}{A}_2 \cdot \underset{\sim}{t} . \tag{84}$$

Obviously, all internal stresses as well as the contact stresses are determined.

Considering a plate under large deflection, which is not in contact with a yielding subgrade, i.e. no friction stresses t will appear, the relations are valid

$$N_r = \bar{N}_r \; , \quad N_\varphi = \bar{N}_\varphi \; ;$$

$$\underset{\sim}{p} = \underset{\sim}{p}_a - \frac{2 \, K_M}{1-\mu} \{ \underset{\sim}{D} \, (2 \, \underset{\sim}{R}_M + \underset{\sim}{I}) - \frac{S_a}{2B} \; [\; \lceil \underset{\sim}{R}_N \cdot \underset{\sim}{\delta}_N \rfloor \; (\underset{\sim}{R}_M + \underset{\sim}{I}) + \qquad (85)$$

$$+ \lceil (\underset{\sim}{R}_N + \underset{\sim}{I}) \; \underset{\sim}{\delta}_N \rfloor \; \underset{\sim}{R}_M] \} \; \underset{\sim}{\delta}_M \; .$$

Considering small deflections, the differential equations (28) hold

$$\underset{\sim}{D} \, (\underset{\sim}{n}_r + \underset{\sim}{n}_\varphi + \underset{\sim}{\Phi}) = \underset{\sim}{\Phi}' - \mu \cdot \underset{\sim}{\Phi}'' \; , \qquad (86.1)$$

$$\underset{\sim}{D} \cdot \underset{\sim}{m} = \underset{\sim}{p} - \underset{\sim}{p}_a + \frac{h}{2} \, \underset{\sim}{\Phi}'' \; . \qquad (86.2)$$

Regarding eq.s (73) and (82), it follows

$$\underset{\sim}{t} = \underset{\sim}{T}^{-1} \cdot \underset{\sim}{D} \, (\bar{\underset{\sim}{n}}_r + \bar{\underset{\sim}{n}}_\varphi) \; , \qquad (87.1)$$

$$\underset{\sim}{p} = \underset{\sim}{p}_a + \underset{\sim}{D} \cdot \underset{\sim}{m} - \frac{h}{2} \, \underset{\sim}{\Phi}'' \qquad (87.2)$$

and furthermore with eq.s (75) and (80)

$$\underset{\sim}{t} = \frac{S_a}{2} \, \underset{\sim}{G} \cdot \underset{\sim}{\delta}_N \, , \tag{88.1}$$

$$\underset{\sim}{p} = \underset{\sim}{p}_a - \frac{2 \, K_M}{1-\mu} \, \underset{\sim}{P} \cdot \underset{\sim}{\delta}_M - \frac{S_a \cdot h}{4 r_a} \, \underset{\sim}{Q} \cdot \underset{\sim}{\delta}_N \, , \tag{88.2}$$

where $\underset{\sim}{G} = \underset{\sim}{T}^{-1} \cdot \underset{\sim}{D} \, (2 \, \underset{\sim}{R}_N + \underset{\sim}{I}) \, ,$

$\underset{\sim}{P} = \underset{\sim}{D} \, (2 \, \underset{\sim}{R}_M + \underset{\sim}{I}) \, ,$

$\underset{\sim}{Q} = \underset{\sim}{A}_2 \cdot \underset{\sim}{G} \, .$

As the differential equation (86.1) does not depend on the normal contact stresses p, and in consequence does not depend on the deflection w, the vector $\underset{\sim}{t}$ of the friction stresses can be determined directly without solving the differential equation. For a central-symmetrically loaded circular plate, the relation between the in-plane stresses σ_r^N and σ_φ^N and the friction stresses t in the interface or between the order of birefringence and the stresses t respectively can be derived by means of the theory of elasticity (fig. 12).

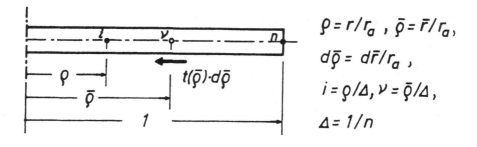

$\rho = r/r_a \, , \, \bar{\rho} = \bar{r}/r_a \, ,$
$d\bar{\rho} = d\bar{r}/r_a \, ,$
$i = \rho/\Delta \, , \, \nu = \bar{\rho}/\Delta \, ,$
$\Delta = 1/n$

Fig. 12: Influence of friction stresses on the stress state
in a central-symmetrical plate

$$\bar{\rho} \leq \rho \leq 1: \ d\sigma_r{}^N = \frac{r_a}{2h} (1-\mu) \ \bar{\rho}^2 \ (\frac{1}{\bar{\rho}^2} - 1) \ t(\bar{\rho}) \ d\bar{\rho} \ ,$$

$$d\sigma_\varphi{}^N = - \frac{r_a}{2h} (1-\mu) \ \bar{\rho} \ (\frac{1}{\bar{\rho}^2} + 1) \ t(\bar{\rho}) \ d\bar{\rho} \ ;$$

$$0 \leq \rho \leq \bar{\rho}: \ d\sigma_r{}^N = d\sigma_\varphi{}^N = - \frac{r_a}{2h} (1-\mu) \ (\frac{1+\mu}{1-\mu} - \bar{\rho}^2) \ t(\bar{\rho}) \ d\bar{\rho} \ .$$

As $\sigma_r{}^N$ and $\sigma_\varphi{}^N$ are the principal stresses, the order of birefringence runs:

$$\bar{\rho} \leq \rho \leq 1: \ d\delta_N(\rho) = \frac{2 \ r_a}{S_a} (1-\mu) \ \frac{\bar{\rho}^2}{\rho^2} \cdot t(\bar{\rho}) \ d\bar{\rho} \ ,$$

$$0 \leq \rho \leq \bar{\rho}: \ d\delta_N(\rho) = 0 \ .$$

Integration yields

$$\delta_N(\rho) = \frac{2 \ r_a}{S_a} (1-\mu) \int_0^\rho \frac{\bar{\rho}^2}{\rho^2} \cdot t(\bar{\rho}) \cdot d\bar{\rho} \ , \tag{89}$$

and after transformation into a finite form

$$(\delta_N)_i = \frac{2 \ r_a}{S_a} (1-\mu) \sum_{\nu=0}^{i} \frac{\nu^2}{i^2} \cdot t_\nu \cdot \Delta \ . \tag{90}$$

Introducing the matrix form, eq. (90) holds

$$\underset{\sim}{\delta}_N = \{ \ (\delta_N)_i \ \} = \frac{2 \ r_a}{S_a} (1-\mu) \ \underset{\sim}{F} \cdot \underset{\sim}{t} \ . \tag{91}$$

As it can be proved the matrix $\underset{\sim}{F}$ to be non-singular, the vector $\underset{\sim}{t}$ runs

$$\underset{\sim}{t} = \frac{S_a}{2r_a(1-\mu)} \; \underset{\sim}{F}^{-1} \cdot \underset{\sim}{\delta}_N \; . \tag{92}$$

This solution may be used instead of eq. (88.1).

Depending on the stiffness ratio between the plate and the subgrade, on the plate radius r_a, on the friction coefficient as well as on the type of external loading, the plate may lift up from the subgrade (fig. 13).

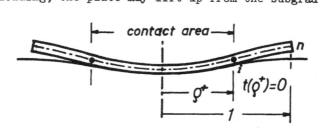

Fig. 13: Determination of the contact area

Under the assumption the radius of the contact area to be r^* (ρ^*), the eq.s (88) and (92) must lead to the result $p(\rho) = 0$, $t(\rho) = 0$, $\rho \geq \rho^*$.

In this case, two areas are to consider, and from eq. (89) follows

$$0 \leq \rho \leq \rho^*: \; \delta_N(\rho) = \frac{2\,r_a}{S_a}\,(1-\mu) \int\limits_0^\rho \frac{\bar\rho^2}{\rho^2}\,t(\bar\rho)\,d\bar\rho \; , \tag{93.1}$$

$$\rho^* \leq \rho \leq 1: \; \delta_N(\rho) = \frac{2\,r_a}{S_a}\,(1-\mu) \int\limits_0^{\rho^*} \frac{\bar\rho^2}{\rho^2}\,t(\bar\rho)\,d\bar\rho = \frac{2r_a}{S_a}\,(1-\mu)\cdot K_1\cdot\frac{1}{\rho^2}. \tag{93.2}$$

<u>Note:</u> For $\rho > \rho^*$ the integral $\int_0^{\rho^*} \bar{\rho}^2 \; t(\bar{\rho}) \; d\bar{\rho} = K_1 = const.$

In the finite form it is valid

$$i < i^*: \; (\delta_N)_i = \frac{2 \, r_a}{S_a} \, (1-\mu) \sum_{\nu=0}^{i} \frac{\nu^2}{i^2} \cdot t_\nu \cdot \Delta \; , \qquad (94.1)$$

$$i \geq i^*: \; (\delta_N)_i = \frac{2 \, r_a}{S_a} \, (1-\mu) \frac{1}{i^2} \sum_{\nu=0}^{i^*} \nu^2 \cdot t_\nu \cdot \Delta \; . \qquad (94.2)$$

But because of unavoidable uncertainties in experimental data and because of finite evaluation methods, the determination of the real contact area may not be of sufficient accuracy. Therefore additional experiments are proposed, e.g. by means of moiré techniques or based on the phenomena of Newton rings. Thus it has been shown, that from photoelastic experiments informations and data can be obtained to determine the in-plane as well as the bending stresses, the contact stresses and the contact area for arbitrarily supported plates of arbitrary geometry considering large deflections. The evaluation of the experimental data should be done by computer programs.

4. Moiré methods

To analyze plates in bending as well as in an in-plane stress state, with large deflection, moiré methods can be applied.

Using the photo-lacquer-technique, a grid of parallel lines with the pitch d is applied to the surface of the plate model. Generally, the distance m of moiré fringes follows from the pitch d_1 and d_2 of the two compared grids:

$$m = \frac{d_2 \cdot d_1}{/d_2 - d_1/} \qquad\qquad (95)$$

Using the double-exposure technique, the recording of the undeformed model (exposure time T/2) yields the reference grid with $d_1 = d$, i.e. the original pitch. Then the plate model is loaded and the deformed grid with pitch d_2 is recorded in the second half of exposure time. According to fig. 14, the pitch d_2 of the deformed grid is influenced by the deflection w, the slope φ and the strain ε of the surface of the plate model.

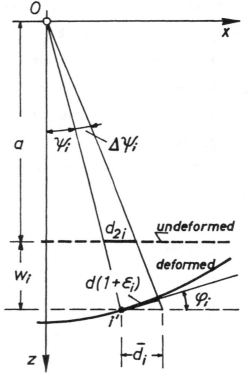

The pitch $d(1 + \varepsilon_i)$ of the deformed grid projected on the plane in the distance w_i parallel to the reference plane of the undeformed model holds

Fig. 14: Influence of surface strains on the moiré effect

$$\bar{d}_i = d(1 + \varepsilon_i) \frac{\cos(\varphi_i - \psi_i - \Delta\psi_i)}{\cos(\psi_i + \Delta\psi_i)} \tag{96}$$

with i, the regarded plate point. As the angle $\Delta\psi_i < \frac{d}{a}$, this angle is approximately set to zero. Then the pitch of the deformed grid projected on the reference plane runs

$$d_{2i} = \bar{d}_i \frac{a}{a + w_i} = d(1 + \varepsilon_i) \frac{\cos(\varphi_i - \psi_i)}{\cos\psi_i} \cdot \frac{a}{a + w_i} \tag{97}$$

Introducing d_{2i} according to eq. (97) and $d_{1i} \triangleq d$ into eq. (95), the distance m of two adjacent moiré fringes influenced by the bending as well as the in-plane stress state can be determined, from which furthermore after some calculations the deflection w_i will be derived:

$$\frac{w_i}{a} = (1 - \frac{d}{m_i})(1 + \varepsilon_i) \cdot \frac{\cos(\varphi_i - \psi_i)}{\cos\psi_i} - 1 , \tag{98}$$

where $m_i = (m_{M,N})_i$ and $\varepsilon_i = (\varepsilon_{M,N})_i$ are depending on the bending and the in-plane stress state.

In a second experiment with the same plate model, a reference grid with the pitch d as the former is projected in a plane wave front on the undeformed plate and recorded in half the exposure time. Then the exposure is finished recording the grid projected on the deformed plate model. In the deformed state, the pitch d_{2i} related to the undeformed state with the pitch $d_{1i} \triangleq d$ runs

$$d_{2i} = \frac{d}{\cos\varphi_i} \cdot \frac{\cos(\varphi_i - \psi_i)}{\cos\psi_i} \cdot \frac{a}{a + w_i} \; , \qquad (99)$$

which yields with regard to eq. (95) the deflection

$$\frac{w_i}{a} = (1 - \frac{d}{m_i}) \; \frac{\cos(\varphi_i - \psi_i)}{\cos\varphi_i \cdot \cos\psi_i} - 1 \; , \qquad (100)$$

where $m_i = (m_M)_i$ depends on the bending stress state only.

As the deflection w_i in both experiments must be equal the strain ε_i at the surface of the model perpendicular to the grid lines is obtained:

$$(\varepsilon_{M,N})_i = \frac{(1 - \frac{d}{(m_M)_i}) \frac{1}{\cos\varphi_i} - (1 - \frac{d}{(m_{M,N})_i})}{1 - \frac{d}{(m_{M,N})_i}} \; . \qquad (101)$$

As φ_i is assumed to be small, $\cos\varphi_i$ is approximately set to 1, and as $d/(m_{M,N})_i$ is very small compared to 1, the denominator may be set to 1 also. Then eq. (101) runs

$$(\varepsilon_{M,N})_i \approx d \; [\; \frac{1}{(m_{M,N})_i} - \frac{1}{(m_M)_i} \;] \; . \qquad (102)$$

Considering the basic principles of optics, the accuracy of such experiments is within the range of 1 μ. In a Cartesian coordinate system, the strains at the upper surface of the model $(\varepsilon_{xo})_i$ and $(\varepsilon_{yo})_i$ are to be determined for perpendicular grid directions, i.e. the grids are to be oriented in the direction of the x-axis and the y-axis respectively. According to the advanced plate theory, the relations are valid (see eq.s (6))

$$\varepsilon_{xo} = n_{,x} + \frac{1}{2} w_{,x}^2 - \frac{h}{2} w_{,xx} \; , \tag{103.1}$$

$$\varepsilon_{yo} = v_{,y} + \frac{1}{2} w_{,y}^2 - \frac{h}{2} w_{,yy} \; . \tag{103.2}$$

The strains in the central plane of the plate then may be calculated according to

$$\varepsilon_x = \varepsilon_{xo} + \frac{h}{2} w_{,xx} \; , \tag{104.1}$$

$$\varepsilon_y = \varepsilon_{yo} + \frac{h}{2} w_{,yy} \; , \tag{104.2}$$

and the deformations according to

$$u = \int \left(\varepsilon_{xo} - \frac{1}{2} w_{,x}^2 + \frac{h}{2} w_{,xx} \right) dx + C_1 \tag{105.1}$$

$$v = \int \left(\varepsilon_{yo} - \frac{1}{2} w_{,y}^2 + \frac{h}{2} w_{,yy} \right) dy + C_2 \tag{105.2}$$

For numerical evaluation, these integral equations are to be transformed into finite formulations.

The values of $w_{,x}$, $w_{,y}$, $w_{,xx}$ and $w_{,yy}$ in the discrete point i are to be derived from the deflections $(w)_i$.

To avoid the well-known disadvantages of derivation of the experimental data instead of discrete values w_i, spline functions $\tilde{w}$ should be
introduced. On the other hand, it is possible also to measure the first
derivatives of the deflection surface w by a third experiment based on
Ligtenberg's[10] method. Neglecting the influence of w_i, the reflection of
the reference grid at the mirrored surface of the plate model (fig. 15)

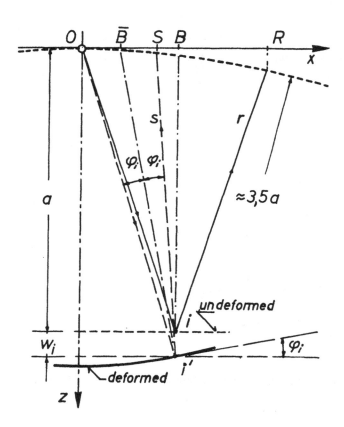

Fig. 15: Principle of Ligtenberg' moiré method

yields the slope φ_x of the deflection surface, if the reference grid is oriented parallel to the y-axis, and φ_y, if the grid lines are oriented in direction of the x-axis.

$$\tan\varphi_x = w,_x = \frac{1}{2a} \left[x_s (1 + \rho) - x_R \right] , \qquad\qquad (106.1)$$

$$\tan\varphi_y = w,_y = \frac{1}{2a} \left[y_s (1 + \rho) - y_R \right] \qquad\qquad (106.2)$$

with $\rho = \dfrac{r - s}{r + s}$.

For large values of a, the distance of the point of observation O from the model surface in the undeformed state and using a cylindrical screen of the reference grid, it can be proved that the factor ρ approximately equals zero, and it follows that

$$\tan\varphi_x = w,_x = \frac{1}{2a} \overline{RS}$$

and $\tan\varphi_y = w,_y$ respectively, where $\overline{RS}$ is the distance between the grid line R observed from O in the undeformed state, and the grid line S observed from O in the deformed state, both reflected in the same point i of the plate model. Superimposing both by photographical recording yields a moiré field, so that each moiré line is an integer multiple of the grid constant d/2a.

$$\tan\varphi_x = w,_x = k_x \frac{d}{2a} , \quad k_x \in \mathbb{N} , \tag{107.1}$$

$$\tan\varphi_y = w,_y = k_y \frac{d}{2a} , \quad k_y \in \mathbb{N} . \tag{107.2}$$

To obtain the second derivatives of the deflection surface, cubic spline functions are introduced for smoothing and balancing the experimental data. Differentiating these spline functions yields the second derivatives of the deflection surface. The second derivatives can also be obtained by the "second moiré" according to Haas/de Loof[11]. Two transparent records of the original first order moiré $w,_x$ are superimposed; one of them is shifted in the direction of the x-axis at Δx. Then new moiré fringes appear, which are lines of constant curvatures:

$$w,_{xx} = (k_{x2} - k_{x1}) \frac{d}{2a\Delta x} . \tag{108}$$

The curvatures $w,_{yy}$ and $w,_{xy}$ are to determine analogous. As the curvatures are determined either by calculation or experiment, the strains in the neutral plane of the plate can be determined according to eq.s (104), and furthermore the in-plane stresses and the related stress resultants N_x, N_y, N_{xy} respectively. The bending as well as the twisting moments are taken from eq.s (12), and the sum of the bending moments M is given also. These results are introduced into the eq.s (27) and yield the contact stresses p and t, as described in chapter 3.

To determine the area, where the plate model remains in contact with

the subgrade, a modified shadow-moiré technique is used. The same plate
model with the applied grid on the plate surface is supported on the subgrade with the grid upside-down.

The surface of the subgrade is prepared to provide sharp and dark
shadows of the grid lines. In parallel light, the moiré fringes are lines
of constant deflection (fig. 16)

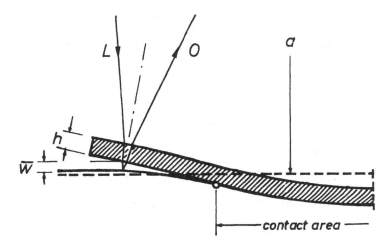

Fig. 16: Determination of the contact area by shadow-moiré technique

$$\bar{w}_i = k\, d\, \frac{a}{\ell}\, \frac{1}{1 - kd/\ell}\,, \qquad k \in \mathbb{N}\,. \tag{109}$$

As $kd/\ell \ll 1$, eq. (109) runs approximately

$$\bar{w}_i \simeq k\, d\, \frac{a}{\ell}\,. \tag{109a}$$

For the whole contact area, $\bar{w}_i = 0$ is valid, and no moiré effect will be observed in that area.

5. Combined method

To avoid some disadvantages of the pure photoelastic method (see chapter 3), and of the pure moiré method (chapter 4), which are mainly caused by the necessity of two models of equal bending and membrane stiffness, these two methods are combined. Then only one plate model is necessary, one surface of which is mirrored by galvano technique for instance.

At first the plate model is analyzed in a Ligtenberg's moiré setup, and as experimental data the slope of the plate surface is obtained according to eq.s (107). These data are evaluated by introducing cubic spline functions, which after derivation yield the curvatures $w,_{xx}$, $w,_{yy}$, $w,_{xy}$ in a Cartesian coordinate system, or $w,_{rr}$, $w,_{\varphi\varphi}$, $w,_{r\varphi}$ in a polar coordinate system respectively. Then also the bending and twisting moments as well as the sum M of the bending moments are determined according to the moment-displacement-relations derived in chapter 2.

Next the same plate model is used for the photoelastic experiment by turning the mirrored surface upside-down and replacing the screen with the master grid by a reflection polariscope in the same experimental set-up as before. Then the order of isochromatic fringes δ and the "characteristic directions" are obtained, as already described in chapter 3.

In general, the principal directions of the simultaneous bending and in-plane stress state do not coincide. The principal directions of the superimposed stress states as a function of z run

$$\tan 2\psi(z) = \frac{2(\tau_{xy}^{M}(z) + \tau_{xy}^{N})}{(\sigma_{x}^{M}(z) - \sigma_{y}^{M}(z)) + (\sigma_{x}^{N} - \sigma_{y}^{N})} \ . \tag{110}$$

Introducing

$$\tau_{xy}^{M}(z) = \frac{12}{h^3} \cdot z \cdot M_{xy} \ ,$$

$$(\sigma_{x}^{M}(z) - \sigma_{y}^{M}(z)) = \frac{12}{h^3} z (M_x - M_y) \ ,$$

$$\tan 2\psi^{M} = \frac{2M_{xy}}{M_x - M_y} \ ,$$

where M_x, M_y, M_{xy} follow from the evaluation of the experimental data by Ligtenberg's moiré method, and

$$\tau_{xy}^{N} = \frac{1}{h} \text{maxN} \cdot \sin 2\psi^{N} \ ,$$

$$\sigma_{x}^{N} - \sigma_{y}^{N} = \frac{2}{h} \text{maxN} \cdot \cos 2\psi^{N} \ ,$$

and furthermore

$$Q = f_1(\text{maxN}, \ \psi^{N}) = \frac{h^2}{6(M_x - M_y)} \cdot \text{maxN} \cdot \cos 2\psi^{N} \tag{111}$$

eq. (110) runs

$$\tan 2\psi(z) = \frac{\tan 2\psi^M \cdot z + Q \cdot \tan 2\psi^N}{z + Q} \tag{112}$$

From this equation, $\cos 2\psi(z)$ follows as a function of maxN and the principal direction ψ^N of the in-plane stress state:

$$\cos 2\psi(z) = \frac{Q + z}{[az^2 + bz + c]^{1/2}} \cdot \tag{113}$$

with the coefficients a, b and c as functions of maxN and ψ^N. The angle $\Delta\psi$ of rotation of the principal axes equals the difference of the principal directions at both the surfaces of the plate model:

$$\Delta\psi = \frac{1}{2} \text{ arc tan } \frac{h \cdot Q \cdot (\tan 2\psi^N - \tan 2\psi^M)}{(Q^2 - \frac{h^2}{4}) + Q^2 \cdot \tan^2 2\psi^N - \frac{h^2}{4} \cdot \tan^2 2\psi^M} \cdot \tag{114}$$

From the photoelastic experiment in a reflection polariscope, the characteristic direction α_1 is obtained, which is half of the angle $\Delta\psi$ of rotation. Thus the angle $\Delta\psi = 2\alpha_1$ is a function of the unknown values of maxN and ψ^N. The order of birefringence is given by the relation

$$\delta = \frac{2}{S} \int_{-h/2}^{+h/2} [(\sigma_x^M(z) - \sigma_y^M(z)) + (\sigma_x^N - \sigma_y^N)] \frac{1}{\cos 2\psi(z)} \, dz \cdot \tag{115}$$

Introducing the expressions of the external forces, eq. (115) holds

$$\delta = \frac{24}{S \cdot h^3} (M_x - M_y) \int_{-h/2}^{+h/2} \frac{z}{\cos 2\psi(z)} \, dz +$$

$$+ \frac{4}{S \cdot h} \cdot \text{max}N \cdot \cos 2\psi^N \int_{-h/2}^{+h/2} \frac{dz}{\cos 2\psi(z)} \, , \qquad (116)$$

where M_x and M_y are already given by Ligtenberg's moiré.

The integrals in eq. (116) can be solved (for solution see Gröbner/ Hofreiter[12]) as follows:

$$\delta = \frac{4}{Sh} \left\{ \frac{6}{h^2} (M_x - M_y) \cdot I_1 + \text{max}N \cdot \cos 2\psi^N \cdot I_2 \right\} . \qquad (117)$$

Thus, the order of birefringence is obtained as a function of maxN and ψ^N. Eq.s (114) and (117) then yield the information on the in-plane stress state maxN and ψ^N. For further evaluation, the relations (52) to (55) are valid.

The results of evaluation are then introduced into eq. (64), which yield the sum of the components of the friction stress t in direction of the coordinate axes. These components are determined by means of the recurrent formula (68). Then the in-plane stress resultants N_x, N_y and the contact stresses p may be calculated, the latter one according to eq.(69).

For given central-symmetrical problems with the only variable r in polar coordinates, Ligtenberg's moiré method yields $\tilde{w}_{,r}$ as a cubic spline function derived from the experimental data given by the moiré fringes. Derivation of the slope function yields $w,_{rr}$ and furthermore the bending moments M_r and M_φ and the sum of the bending moments $M = - B(\tilde{w},_{rr} + \frac{1}{2} w,_r)$.

The photoelastic experiment yields $\delta(r)$, the curve of the birefringence order over r, and evaluation according to eq.s (73) and (28) the in-plane stresses σ_r^N, σ_φ^N and their resultants N_r, N_φ respectively, as well as the contact stresses t and p.

It is obvious, that the combined method is less sophisticated than those described in chapters 3 and 4, less time-consuming and easier to handle. In consequence, this method is less sensitive to errors and un-certainties.

6. Strain gage method

To measure the mechanical strains electrically by means of strain gages seems to be the most reliable method (see Rohrbach[13]). But in con-tradiction to the optical methods, informations are only given in those points, where strain gages are applied to the model and - although by the recent technical progress the gage length has become very short - only average values with reference to the length of the strain gage are ob-tained. An additional problem appears regarding plates in contact with respect to the application of strain gages in the interface between the

plate model and the subgrade. The strain gages should not have any contact with the subgrade.

The principle of actual measurements is the "Wheatstone's bridge".
To analyze plate problems, the strain gages are to bridge across in a
complete external circuit. If the actual strain gages and the gages TK
for temperature compensation are arranged in a complete bridge circuit,
as shown in fig. 17a, the measured quantity is equivalent to $\varepsilon(1) + \varepsilon(2)$;
an arrangement according to fig. 17b yields the quantity $\varepsilon(1) - \varepsilon(2)$.

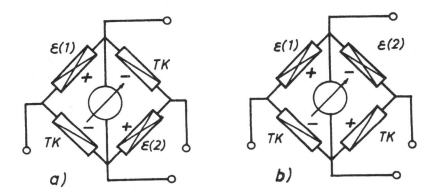

Fig. 17: Wheatstone bridge circuit

To consider the Poisson's ratio μ already in the measuring process,
a potentiometer is put into the circuit as an additional ohmic resistance
to control the current in the active strain gage (fig. 18).

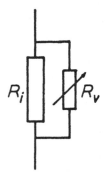

According to Kirchhoff's law for parallel con-
ductors, the relation is valid

$$\frac{1}{R} = \frac{1}{R_i} + \frac{1}{R_v} \, , \qquad (118)$$

Fig. 18: Ohmic resistance
with potentiome-
ter controller

where R_i denotes the ohmic resistance of the
active strain gage, R_v that of the control resistor.

As the condition

$$R = \mu \cdot R_i \qquad\qquad\qquad (119)$$

must be satisfied, the resistance of the potentiometer controller is to

switch to

$$R_v = \frac{\mu}{1+\mu} R_i \, . \qquad\qquad (120)$$

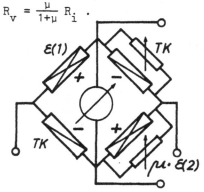

Then the circuit (fig. 19) yields
the measured value

$$r = \varepsilon(1) + \mu \cdot \varepsilon(2) \, . \qquad (121)$$

Fig. 19: Potentiometer controller
in a bridge circuit

Based on the preceding considera-

tions, strain gage rosettes are applied to opposite points on both surfaces of the plate or the plate model to be investigated. These rosettes
are bridged across together with the control resistor. The output signals
of such circuits are proportional to the internal forces, as are the
bending and twisting moments of the bending stress state and the resultants of the in-plane stress state.

Under the supposition Hooke's law of elasticity to be valid, the
relations between the internal forces and the strains or the measured
values respectively run (see Laermann[2]):

$$M_x = - \frac{B}{h} [(\varepsilon_{xo} - \varepsilon_{xn}) + \mu (\varepsilon_{yo} - \varepsilon_{yn})] \tag{122.1}$$

$$M_y = - \frac{B}{h} [(\varepsilon_{yo} - \varepsilon_{yn}) + \mu (\varepsilon_{xo} - \varepsilon_{xn})] \tag{122.2}$$

$$M_{xy} = - \frac{B}{h} (1 - \mu) \{ (\varepsilon_{x+45^\circ,o} - \varepsilon_{x+45^\circ,n}) -$$
$$- \frac{1}{2} [(\varepsilon_{xo} - \varepsilon_{xn}) + (\varepsilon_{yo} - \varepsilon_{yn})] \} \tag{122.3}$$

$$N_x = \frac{E}{1-\mu^2} \cdot \frac{h}{2} [(\varepsilon_{xo} + \varepsilon_{xn}) + \mu (\varepsilon_{yo} + \varepsilon_{yn})] \tag{122.4}$$

$$N_y = \frac{E}{1-\mu^2} \cdot \frac{h}{2} [(\varepsilon_{yo} + \varepsilon_{yn}) + \mu (\varepsilon_{xo} + \varepsilon_{xn})] \tag{122.5}$$

$$N_{xy} = \frac{E}{1+\mu} \cdot \frac{h}{2} \{ (\varepsilon_{x+45^\circ,o} + \varepsilon_{x+45^\circ,n}) -$$
$$- \frac{1}{2} [(\varepsilon_{xo} + \varepsilon_{xn}) + (\varepsilon_{yo} + \varepsilon_{yn})] \} . \tag{122.6}$$

The sum M of the bending moments follows from eq.s (122.1) and (122.2)

$$M = - \frac{B}{h} \left[(\varepsilon_{xo} - \varepsilon_{xn}) + (\varepsilon_{yo} - \varepsilon_{yn}) \right] , \qquad (123)$$

and the curvatures are taken from the data:

$$w,_{xx} = \frac{1}{h} (\varepsilon_{xo} - \varepsilon_{xn}) , \qquad (124.1)$$

$$w,_{yy} = \frac{1}{h} (\varepsilon_{yo} - \varepsilon_{yn}) ; \qquad (124.2)$$

$$w,_{xy} = - \frac{M_{xy}}{B(1-\mu)} . \qquad (124.3)$$

Introducing these data into the finite differential equations (27), yields the contact stresses p and t in those discrete points i, where strain gage rosettes are applied. It must be pointed out, that in contradiction to the afore-described optical methods, strain gage measurement yields no full-field informations but only informations in small numbers of discrete points. On the other hand, there are less restrictions with respect to the model materials, which can be used.

Because of the recent developments in automatic data acquisition and data processing, the signals of the experimental process are put on-line into a multiposition measuring device with automatic computer-operated control of the experimental as well as of the measuring and evaluation process (fig. 20).

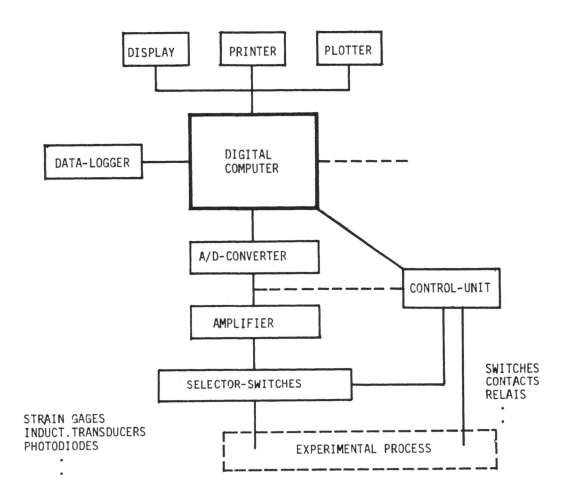

Fig. 20: Block diagram of a multiposition measuring arrangement

7. Final remarks

An advanced plate theory has been described to consider large deflections as well as plates in contact with a yielding subgrade. In consequence, the most applied experimental methods as photoelasticity, moiré
techniques and the strain gage method are adjusted to this advanced theory. Also a combined method has been described to analyze the simultaneous
bending and in-plane stress state experimentally. The experimental data
are introduced into the differential equations, which describe the regarded problem by theory. With the experimental input data the solution
of the constitutive equations is then very easy, if they are transformed
into a finite form. All considerations are still based on the assumption
of Hooke's law of elasticity to be valid. It must be mentioned, that the
final results, i.e. the stress state in the plate as well as the contact
stresses, are to transmit from the model to the prototype. Therefore,
proper transmission formulas must be considered, which are based on the
principles of similarity.

8. References

1. Wolmir, A.S., *Biegsame Platten und Schalen,* VEB-Verlag f. Bauwesen,
   Berlin,1962.

2. Laermann, K.H., *Experimentelle Plattenuntersuchungen - Theoretische
   Grundlagen -* , Werner-Verlag, Düsseldorf, 1972.

3. Frocht, M.M., *Photoelasticity*, John Wiley & Sons Inc., New York, Vol. I (1941), Vol. II (1948).

4. Wolf, H., *Spannungsoptik*, Bd. 1, 2. Aufl., Springer-Verlag, Berlin, Heidelberg, New York, 1976.

5. Drucker, D.C. and Mindlin, R.D., Stress analysis by three-dimensional photoelastic methods, *Journ. Appl. Phys. 11*, 1940.

6. Kuske, A., *Einführung in die Spannungsoptik*, Wissenschaftl. Verlagsgesellschaft, Stuttgart, 1959.

7. Aben, H., Optical phenomena in photoelastic models by the rotation of principal axes, *Experim. Mechanics 6*, 1966.

8. Haberland, G., Einige Auswerteverfahren bei spannungsoptischen Plattenuntersuchungen, *Ing. Arch. 30*, 1961.

9. Laermann, K.H., Ein Auswerteverfahren für die spannungsoptische Untersuchung von Platten, bezogen auf Polarkoordinaten, *Die Bautechnik 44*, 1967.

10. Ligtenberg, F.K., Over een methode om door een eenvouding experiment de momenten in stijve platen te bepalen, *De Ingenieur 64*, 1952.

11. de Haas, H.M. and Loof, H.W., An optical method to facilitate the interpretation of moiré-pictures, *Experimentelle Spannungsanalyse, VDI-Berichte Nr. 102*, VDI-Verlag, Düsseldorf, 1966.

12. Gröbner, W. and Hofreiter, N., *Integraltafel*, 5. Aufl., Springer-Verlag, Wien, New York, 1975.

13. Rohrbach, Chr., *Handbuch für elektr. Messen mechanischer Größen*, VDI-Verlag, Düsseldorf, 1967.

FOUNDATIONS OF EXPERIMENTAL MECHANICS:
PRINCIPLES OF MODELLING,
OBSERVATION AND EXPERIMENTATION

Jerzy T. Pindera
Faculty of Engineering
University of Waterloo
Waterloo, Ontario, Canada N2L 3G1

PREFACE

As it is indicated in the preface to the book, the major problem of contemporary experimental research in mechanics is the growth of discrepancies or even contradictions between the theory and technique of producing, collecting and processing of information-carrying signals, and the typical theoretical bases for methods used in Experimental Mechanics.

Several more or less extensively discussed reasons have contributed to a rapid development of the theory and technology of contemporary measurement methods and to unexpectedly high precision and resolution of measurement techniques presently available; the conceptual influence of modern physics and of requirements of modern industry is unmistakable.

However, the development of theories and related methods and tech-

niques of experimental research in mechanics apparently has not kept step
with the above development. As a result, at present it is easy to obtain
highly accurate results, the reliability of which is often questionable.
This often leads to undesirable theoretical and technological consequences.
Moreover, it happens that the presented experimental evidence is supported
by predictions of an incorrectly chosen theory.

It appears that the main reason for these discrepancies is conceptual.
Traditionally, mechanis has been concerned with "discovering of laws" and
"determination of properties". It has been believed that the "discovered
laws" and the "discovered constitutive equations" can describe and rigor-
ously predict the behaviour of real systems and materials. Obviously,
within such a frame of reference a "governing equation" could not be
"invented"; to be "true", it had to be "discovered".

The time when the theory of elasticity, for instance, was rigorously
called "the mathematical theory of elasticity", a term clearly indicating
application of concept of a mathematical universe, has unfortunately
passed. It is still customary to assess some new experimental methods in
mechanics by comparing experimental data with questionable numerical pre-
dictions of too simplified a theory. Fruitful concepts related to the
modelling of physical reality, including concepts of invented mathematical
models, are obviously incompatible with concepts of discovered laws or
discovered constitutive equations, as is the concept of a useful paradigm.

In my opinion, a need has arisen to discuss the theoretical found-
ations of the classical and emerging approaches in Experimental Mechanics
and to outline the major theoretical, methodological and technological

*problems. In this vein the main topic of this work is the methodology of
development of sufficiently comprehensive physical and mathematical models
of total actual systems, including the experimenter. It is evident that,
e.g., within the presented frame of reference the term "physical model"
denoting a conceptual physical model obviously is not identical with the
terms "scale model" or "inconic model".*

*I am using the term "physicalisation of mechanics" to denote the chosen
approach. This approach could be extrapolated to encompass the growing
awareness of the long-term ecological effects produced by technology.*

*The usefulness of less phenomenological and more physical approach is
illustrated by examples. The particular topics discussed or outlined are:*

*(a) New trends in contemporary approaches to experimental research in
mechanics based on more comprehensive modelling of real processes and
involved systems, in particular:*

*- Development of a unified and multi-discipline approach in experimental
mechanics: principles of design and evaluation of the conceptual physical
and mathematical models of behaviour of materials under loads, and of the
responses of the testing and measurement systems.*

*(b) Need for an extension of theoretical basis for analytical and experi-
mental determination of strain and stress states, caused by recent advances
in experimental techniques:*

- Improvement of some mathematical models of stress state:

- Consideration of the actual material responses to loads:

*- Introduction of more comprehensive models of interaction between energy
flow and matter, especially with regard to radiant energy, and to the*

actual patterns of energy flow and transfer.

The examples illustrating the major points pertain to some classical and new experimental techniques requiring a more comprehensive basis:

- Dynamic responses of engineering materials: thermoelastic effect in solids and its engineering significance.

- Photoelasticity: spectral birefringence in deformed solids (creep, recovery, isochronous presentation); spectral birefringence in liquids, scattered light measurements; stationary and scanning techniques; new scattered light techniques; utilization of the isodyne fields; transfer functions of photoelastic systems.

- Limits of application of some classical analytical solutions for stress evaluation.

- Ultrasonic methods: limits of applicability of geometric optics.

Rational development of physical, mathematical and iconic models of materials and processes constituting foundations of Experimental Mechanics is the underlying theme of this work.

As the main subject is the methodology of choosing or developing the most commensurate theory for a given engineering task, a problem which cannot be presented as a continuous logical and unequivocal mathematical proof, I have adapted the lecture format for the structure of this work.

I am indebted to my son, Maciej-Zenon Pindera, for fruitful discussions when editing the rough draft.

November, 1980.

PART I: CHOICE OF METHODOLOGY OF
EXPERIMENTAL RESEARCH IN MECHANICS

1. INTRODUCTION

Motto: 'It is true that
     "Nothing is more practical than a theory"
     Provided - however -
     That the assumptions on which the theory is founded
     Are well understood.
   - But, indeed,
     Nothing can be more disastrous than a theory
     When applied to a real problem
     Outside of the practical limits of the assumptions made,
     Because of an homonymous identity
     With the problem under consideration.'
                       (from own experience)

The major patterns of research in various fields of applied mechanics
have in the past been simple and enjoyable. A set of assumptions could
have been made, mostly intuitively, i.e., primarily for the reasons of
convenience, and the predictions of the evaluated functional relationships
could have been compared directly with the chosen experimental data which
were supposed to truly represent the relationships between the physical
quantities of interest. The scattering of experimental data was usually
related to the imperfections of measurement apparatus: consequently, a
suitable statistical treatment of experimentally obtained data was assumed
to be sufficient to determine the "true event".

To predict the behaviour of real bodies under real conditions, it has
been customary to apply the solutions derived for hypothetical bodies and

hypothetical conditions on the basis of "intuitive" analysis. For instance, it has not been deemed necessary to accept the concept of multiparameter responses of bodies to particular flow of energy, when the concept of practically constant "properties" of materials worked so satisfactorily.

A strong tendency has existed to believe that a rigorous solution found for an assumed body under assumed conditions also represents a rigorous solution for a real body. The term "idealization" has been understood literally, as a process of discovering the true ideal properties of bodies or systems.

Such an approach caused the analysts to overlook the fact that a rigorous solution for a model of a real body represents very often a crude empirical solution for the body itself, in spite of the homonymous identity resulting from identity of labels attached to a body and to a simplified model of it.

As a result, the patterns of a significant part of research in mechanics have been deviating more and more from the patterns of research in pure sciences. The approach has been becoming more and more phenomenological and technological. Very often new analytical solutions have been supported by experimental evidence collected in a manner which would not satisfy the requirements typical in industrial testing. Also, the same terms have been used to denote different things, as, for instance, the term "physical model". The belief has been growing that it is possible to investigate the real relationship between parameters of real physical processes on the basis of a very simple understanding of real phenomena, i.e., on the basis of a very simple picture of nature. It has happened too often that the already acquired knowledge was lost or replaced by very crude theories.

Research in mechanics related to the fantastic technological developments in the fields of the aerospace and nuclear industries became very sophisticated and, as a result, somewhat separated from the general level of experimental research in mechanics, in two major aspects: conceptual and technological. The requirements of the leading branches of modern technology have enforced a necessary progress in the related experimental research. However, in the less demanding branches of industry which are using less advanced technology and are more constrained financially, a more elementary level of research is still prevailing. The differences in experimental techniques result mainly from financial factors; the differences at the conceptual level however are rooted partly in some psychological factors, partly in the not up-to-date university curricula, and partly in a noncritical acceptance of the concept of useful paradigms.

In addition, fundamental discrepancies have been developing between the progress in the theory and technique of collecting and process of signals of various origins, and the rate of development of theories used to collect, process, evaluate and assess the experimental data.

The rapid progress in the information theory, control theory, computer science, materials science and electronics has led to the development of new generations of testing, measurement and recording systems. Those systems are of very high reliability and accuracy. It is easy to measure, display and record the time periods in the nanosecond range, the electric signals in the microvolt range, the force in the micronewton range, the

temperature differences with the accuracy of 0.001°C, etc. The measurement, analysis, transformation from one domain into another, display and
recording of time-dependent signals are easy, reliable and inexpensive in
the range from DC to hundreds megahertz. In addition, the introduction of
microprocessors led to automation of tedious and boring operations, such
as adjusting of measurement systems, Fourier analysis, statistical evaluation, differentiation and integration of recorded patterns, final evaluation of results, presentation of results as functions of chosen parameters,
etc.

As a result of this development, it is at present easy, fast and relatively inexpensive, to convert any physical quantity into an electric
signal, to process it in a desired manner, to display it for an arbitrary
period of time, to analyze it and to record it in a chosen form. The reliability and accuracy of these operations can be very high, depending on
the needs. In other words, the major problems of every experimenting
engineer and researcher - namely, the resolution of the measurement system
and the accuracy of processed data - ceased to exist.

However, a new problem has arisen: the problem of reliability and
accuracy of results of research and measurement. This problem is directly
related to the correct or incorrect choice - or development - of the conceptual physical and mathematical models of the event of interest; it is
also directly related to the design and theory of experiments chosen to
investigate that event.

It can be noted that several more or less serious discrepanices are
growing between the levels of the theory of experimentation and theory and

technique of measurement.   Similar discrepancies are growing between the theoretical backgrounds chosen as a foundation of various experimental methods and techniques, even in cases when these methods are closely related.   In spite of the available information, quite often an experiment is based on too primitive a theory, when a more comprehensive theory could be easily applied to increase the reliability of results, and - eventually - to establish new, more comprehensive relationships.

As a result, a very unsatisfactory situation has developed in experimental mechanics regarding the existence and acceptance of minor and major discrepancies on the conceptual level of treatment of the media, fields and processes;   the problems involved are both of the methodological and theoretical natures.   A more basic approach and the resulting concepts, such as the concept of a new or useful paradigm, cannot be incorporated into the usually accepted frame of reference.   For instance, it happens too often that the inaccurate or even incorrect predictions of some analytical solutions for the stress states are taken as a measure of correctness of the obtained experimental data.

Since several years a new approach is developing in various fields of mechanics and of applied mechanics in particular; a new language is being introduced for a more commensurate description of the processes being studied, and curricula at the leading universities are being changed accordingly.

It seems that these rapid changes which are bringing the various fields of applied mechanics closer - in approach - to natural sciences and to foundations of scientific understanding of reality, were also triggered -

or enforced - rather indirectly by the progress made in the fields of
information and control theory. Such a pragmatic process seems to be
caused by the necessity and usefulness, rather than by the fundamental
intellectual reasons.

It also seems that one particular major factor is contributing to this
process of - as one could call it - physicalisation of modern applied mechanics, namely the changing sense of values. In the engineering sciences,
the centre of gravity is shifting from the narrow, mainly economical-
technological short-term criteria towards the general long-term social-
ecological-economical-technological criteria.

This new attitude automatically enforces thinking in terms of systems,
of interaction between system components, of patterns of flow of energy
and information, of the physical sense of every theory. It enforces the
acceptance of the causal-stochastic character of all phenomena and, consequently, of all real relationships and responses.

In this approach the observer - research, design engineer - represents
a component of a larger system; this component can be described more or
less adequately by a conceptual physical model which, in turn, can be
reduced to a set of transfer functions and impedances. When the patterns
of information flow to the human mind are understood as the patterns of
response of human senses and mind to the various signals, then the terms
like "illusion" or "misleading information" ought to be replaced by the
term "response to signals" which can be influenced by the signals of
interest and by various modifying and noise signals.

Of course, a clear assumption must be made regarding the significance

and the meaning of observations, of conclusion, of the accepted theories,
etc. Lack of clarity regarding these problems can and too often does
result in unexpected but logical and foreseeable failures of engineering
structures or processes. The motto formulated earlier, attempts to present this issue in a nutshell. There is an obvious intellectual and
economic need for a more rigorous approach to all the involved problems.

With regard to the experimental research in mechanics of solids, a
need is growing for development of more comprehensive physical and mathematical models of responses to various loads of materials and systems used
in experimental analysis of behaviour of materials, components and structures. This process can lead to a fruitful cross-pollination between all
fields of applied mechanics such as analytical mechanics, numerical mechanics, experimental mechanics and the mechanics of materials. To do this
effectively, to successfully apply the developed relationships in design,
and to optimize the design according to the requirements of modern economy,
it seems necessary to acquire an adequate understanding of the following
issues:

- Methods of approach in experimental mechanics: simple deterministic
phenomenological approach; causal-stochastic physical approach.

- Foundations of theories of experimentations, measurements and evaluation of results: flow of energy and information through the testing,
measuring and recording systems; systems per se; transfer functions;
impedances as measures for energy transfer.

- Modelling in mechanics: process of development of the conceptual
physical, mathematical and iconic models of responses of real bodies and

real structures;   choice of major parameters;   reliability of models.

- Coupled responses of real bodies to various forms of energy inputs:
energy flow as a detector of responses of materials and systems; rheological responses of materials used in model mechanics to mechanical loads
and to temperature: mechanical responses, optical responses;   influence
of the kind and the form of energy: spectral birefringence, spectral dependence of principal axes;   relations between the viscoelastic deformation
and the spectral birefringence;   linear ranges of response.

- Limitations of the typical mathematical models of material responses
(constitutive equations);   limits of applicability of geometric optics
in ultrasonics and photoelasticity.

There still exists a belief that a rigorous solution of a constitutive
equation represents a rigorous solution of the real physical problem. It
must be remembered that the mathematical methods can yield rigorous results
only when applied to mathematical problems. Mathematics, which is often
denoted as a "man-made universe", investigates behaviour of logical structures which are not necessarily commensurate with reality. A rigorous
solution of a differential equation can represent a very poor solution of
a homonymously identical engineering problem.

The aforementioned issues can be illustrated by a statement made by
a leading scientist who "estimates that half or more of the numerical data
published by scientists in their journal articles are unusable because there
is no evidence that the researcher actually measured what he thought he was
measuring or no evidence that possible sources of error were eliminated

or accounted for".[*]

A more detailed presentation and discussion of the underlying theories and approaches are given in References [1-19], and in modern textbooks on physics and materials science, or on designing,[102-103].

The topic reviewed above can be presented and analysed in many ways. The objective of the chosen presentation is to accentuate the significance of basic axioms and basic theoretical foundations.

2. CONTEMPORARY DEVELOPMENTS IN EXPERIMENTAL MECHANICS

The amount of scientific and technological information increases so rapidly that it is not possible any more to have all of the available facts and information concerning a particular development in a particular field.

Because of this process and in oider to be able to keep step with the scientific and technological progress, it is necessary to put less emphasis on details and more emphasis on the knowledge of basic scientific facts and on the understanding of the methods leading to production of scientific and technological information.

Understanding of the methodology and the foundations of contemporary engineering research and development is necessary for a correct application of the knowledge acquired. Statements such as "I do not care where my formula came from;— as an engineer I only want to know how to use it", when accepted, produce unnecessary and expensive failures of structures, machines and systems. From this point of view it is desirable to critically review

[*] Quoted by DEAN, R.C., Jr., in "Truth in Publication", *Transactions of the ASME*, June, 1977, p. 270.

the meaning of all the basic terms, definitions and assumptions of
mechanics, some of which often are only tacitly accepted.

2.1   General Characteristics

Considering the chosen methods and techniques of investigations,
applied mechanics can be roughly presented as a system that consists of
three closely related fields:

- Analytical Mechanics;

- Numerical Mechanics;

- Experimental Mechanics.

The meaning of the first two terms is self-evident. The meaning of
the term "Experimental Mechanics" is changing. The methods and techniques
developed and applied in experimental mechanics are coming closer to
physical research and methodology of materials science – and are getting
farther from technological testing.

The rapid and successful development of numerical methods and the less
spectacular but also essential developments in analytical mechanics have
led to a decrease of the significance of experimental stress analysis as
one of the major areas of traditional experimental mechanics. These developments, strongly influenced by demands of contemporary industry, have also
led to a rapidly increasing demand for experimental assessment and/or veri-
fication of mathematical models used as a basis for analytical and numerical
solution of stress states and of behaviour of materials and structures.
Demands of industry have also enforced expansion of the theoretical basis
of experimental research in mechanics, and – consequently – have led to the
development of a new, strong trend called here "physicalisation of

mechanics". This process is not consistent. However - as a result - in the present stage of development, Experimental Mechanics already can be understood as a developing system of theories, methods and techniques, which:

- is based on experimental physics and consequently accepts and applies the contemporary approaches, methods and techniques of natural sciences;

- uses concepts, methods, techniques and solutions developed by analytical and numerical mechanics, materials science, information and control theories;

- contributes to development of engineering theories;

- serves engineering research and engineering practice.

The explicit and implicit meaning of this definition is discussed further in this text.

2.2 Approaches: Assumptions, Principles, Concepts, Axioms

New trends in experimental research are founded on the following understanding of reality, made explicitly or tacitly:

- physical phenomena do exist;

- human mind is able to design and further develop commensurate models of the real physical phenomena;

- to obtain any reliable information on the reality it is necessary to produce and to analyse a suitable energy flow (i.e., a speculative approach is not sufficient);

- any energy flow alters the event under observation.

The following set of principles, concepts and axioms is taken as a basis of modern analytical and experimental methods and techniques in engineering mechanics.

The principles of: (a) causality and (b) correspondence. The concepts of: (a) mass, (b) energy and (c) system. The experimentally established axioms: (a) the principle of conservation and equivalence of mass
and energy, and (b) the three laws of thermodynamics.

Very often mass is considered a manifested quantity that represents
inertia. Thus intertia may be taken to be a fundamental quantity which
characterizes all material bodies.

Information on behaviour of real bodies is produced, detected and
carried by a suitable form of energy. It follows from the first law of
thermodynamics that it is theoretically impossible to obtain any information without expenditure of energy; consequently it must be accepted
that in experimental research the observer must always influence the
process being investigated.

Since any disturbance in real bodies is inseparably connected with
flow of energy, the velocity of propagation of a particular form of
energy is the velocity of the related real perturbance.

Obviously, any reliable solution in applied mechanics must comply
with both the above principles - causality and correspondence - and with
the related conclusions.

Within this framework, which is not as self-evident as it may appear
and therefore is not always accepted totally, one can choose various
approaches to the understanding and description of the reality.

It is obvious that the approach chosen in research, analytical,
numerical or experimental, influences the intellectual and the technical

values of evaluated relationships and determines their efficacy in
engineering and technology.

The variety of common approaches can be represented by two extreme
cases:

(1) The phenomenological, simple-deterministic approach; nature (reality)
is ideal and simple deterministic; only the human mind is imperfect and
distorts the beautiful unequivocal relationships between the actual behaviour of physical bodies, which can be described in terms of inherent
properties; a certain cause produces a certain effect only; the predict-
ability of behaviour of a body can be extended arbitrarily in space and in
time with desired accuracy.

(2) The physical, causal-stochastic approach: the physical phenomena are
complicated, inter-related, and depend on deterministic and nondeterministic
relationships; a certain cause may produce a certain effect, with certain
probabilities and certain delays; the accuracy and the reliability of predictions decreases with the increasing spatial and temporal distances.

In practical application the phenomenological approach leads to an
adequate description of an actual event, when based on a feasible mechanism
of phenomenon, whereas the physical approach requires that the actual
mechanism of phenomenon be understood.

2.3 Model Development in the Phenomenological and Physical Approaches

A. *In the phenomenological, simple-deterministic approach* it is assumed
that it is possible to isolate a real phenomenon, to simplify it in an
unequivocal manner, and to describe it - more or less rigorously - by
simple causal relationships. "Simplification" is often believed to be

synonymous with "idealisation".

It is assumed that the real phenomena can be modelled – in mind, or in matter and energy – as "true models", that they can be described commensurably by ordinary differential equations, and that they can be observed and studied "as they are".

It is also assumed, openly or tacitly, that it is possible to obtain information on the observed phenomenon without disturbing it, and that the process of producing and collecting information is not necessarily related to thermodynamics.

Consequently it is assumed that in experimental studies the noise in measurement systems can be eliminated, that the measurement systems can truly reproduce the amplitude and the phase of the measured quantity, and that the scattering of measurement is a typical stochastic process resulting from imperfections, and can be minimized arbitrarily.

Hence, according to the phenomenological approach:

– the statistical (e.g., mean value) curve in Figure 1 is unequivocal and represents truly the physical process under study, when a systematic error is eliminated;

– a conceptual model or an iconic model of a real phenomenon can be designed as a "true" model: consequently any "correct model", for instance a conceptual model or a scale model made of matter and energy, can represent the related phenomenon in a unique manner;

– the prediction of a model represents a solution of a problem;

– qualitatively different forms of response of real bodies are unrelated, consequently the ordinary differential equations are suitable

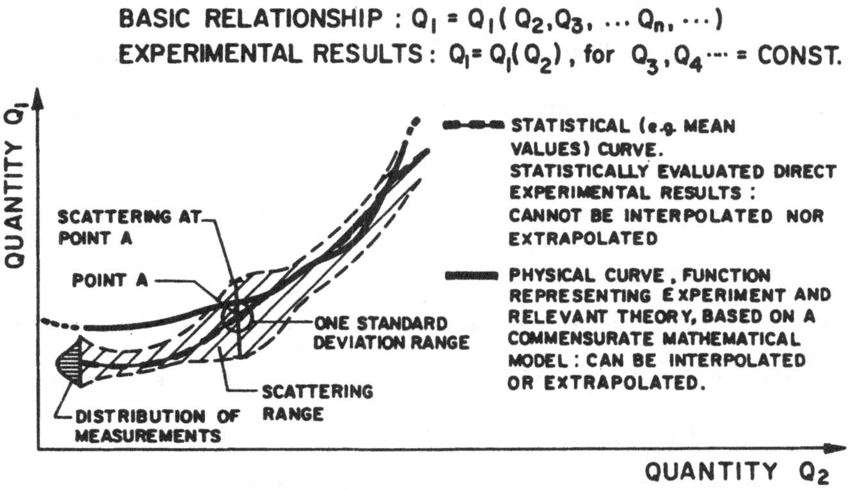

BASIC RELATIONSHIP : $Q_1 = Q_1(Q_2, Q_3, \ldots Q_n, \cdots)$

EXPERIMENTAL RESULTS : $Q_1 = Q_1(Q_2)$, for $Q_3, Q_4 \cdots = $ CONST.

Figure 1 - Principles öf interpretation of experimental results:
phenomenological (purely empirical), and physical interpretations. Broken line represents measurements.
Continuous line represents actual relationship.

for rigorous representation of real bodies;

- a rigorous solution of a model, e.g., of constitutive equations,

represents a rigorous solution of related physical problem;

- the Model Mechanics may and should be based on the concept of rigorous

"Model Laws", which can describe rigorously the relations between a real

phenomenon and its real model;

- the behaviour of real materials can and should be based on the concept

of the intrinsic "properties of materials", since physical phenomena can

be unequivocally and rigorously presented in the form of linear different-

ial equations, the coefficients of which represent inherent properties of

materials or systems;

- the human mind is able to perceive the physical phenomena "as they

are"; deviation from this rule is called "illusion";

- an observer is not a part of a measuring system;

- the "laws of Mechanics" are discovered;

- it is possible to obtain information without changing the entropy of a system;

- emphasis is put on "accuracy" of measurements as described by precision of measurement of the signal of interest;

- measurement process is understood as a process of collecting and transforming of undisturbed information about undisturbed process; consequently the concept of "measuring chain" may be introduced.

B. *In the physical, causal-stochastic approach* it is assumed that it is not possible to isolate any real phenomenon. There is always an interaction between the phenomenon under study and the surroundings, including the observer.

"Simplification" means simplification — there is nothing "ideal" in simplicity. The so-called "ideal" or "true" models are but crude models of reality, which however can be very useful when their limitations are known.

Some phenomena cannot be described by causal relationships; the noise level, as well as the scatter of results, cannot be minimized arbitrarily.

Because of the nature of real phenomena, partial differential equations are more suitable for a commensurate presentation of fundamental relationships; an ordinary differential equation is always an approximation which neglects the influence of all but one of the infinite number of material and systems variables.

In this approach, the degree of rigorousness to which a problem is solved analytically depends not only on the rigorousness of mathematical transformations and solution; it depends also on the rigorousness of the proofs which justify the assumptions made and justify all the physical and mathematical simplification made. For instance, depending on values of major material and system parameters, the same analytical solution for a vibrating beam can represent a rigorous solution, or it can represent a crude empirical formula if, for example, the process under study is assumed to be isothermal but the actual material exhibits a large thermoelastic effect.

Since any information about real phenomena is produced, collected and represented by particular forms of energy, it is recognized that any observation influences the observed process: the process of collecting information is a process subject to the laws of thermodynamics. Because – according to the physical approach – it is theoretically not possible to observe any phenomenon "as it is", the mean value curve in Figure 1 represents the relation between two physical quantities Q_1 and Q_2 influenced by loading and measuring process, both quantitatively and qualitatively. Thus, such a relation is pertinent only to a given experiment, and cannot be properly utilized without reliable information on the theory of experiment. According to the physical approach:

- In Figure 1, the mean value represents results of the performed experiment (measurement) only.

- Scatter of results is caused by imperfections of measurements and by scattering of all physical processes involved.

- Models of real phenomena are always simplified, regardless of whether they are conceptual models or iconic models; iconic models are not "real" models. A model is always an approximation that should reproduce only some selected patterns of a given phenomenon, in a distorted manner. Obviously, it is possible to design many conceptual and iconic models of the same phenomenon, depending on the chosen criteria of modelling, as may be seen from the similarity theorem. For instance, some manufactured iconic models of the same structural component may be geometrically quite dissimilar.

- The prediction of a model represents *directly* only the prediction of a model; it does not represent *directly* a solution of a problem, although it may be related to it.

- The above facts lead to the concept of "degree of correlation" between a real phenomenon and the models of it, and to the concept of "correlation laws".

- The term "property" in the sense of an intrinsic quantity of a structure or a material is replaced by the term "behaviour" or "response". In this context the term "material property" is understood in the sense of a manifested quantity only, within the framework of an assumed conceptual model, as a parameter of this model only, and may be referred to as a material response parameter.

- Rigorous solution for a model can be a very crude and unreliable solution for a real problem.

- Fundamental "laws of Mechanics" are recognized as axioms; that is, they are invented.

- A measuring process is understood as a process of producing, collecting, transforming and recording of information of interest which is represented by a particular form of energy; consequently it is accepted that it is impossible to obtain an undisturbed information and to process it without distortion. This leads to the concept of a measuring system which is characterized by responses and can be described by transfer functions and impedances. A measuring system is a component of an information producing and collecting system.

- The observer is always a part of a measuring system.

- The human mind and senses, being a part of nature, perceive the physical phenomena and transform the obtained information in a phenomenologically similar manner as the measuring instruments; consequently, the term "illusion" is replaced by the term "response of human mind and senses" or "human transfer function".

Emphasis is put on the "reliability" of experimental data, which depends on the degree of correlation between the phenomenon itself and the accepted model of it.

Experimental results are treated and understood in the manner depicted in Figure 1 by the physical curve.

A specific example is given in Figure 2. The discrepancies between the indicated (recorded) temperature alteration of a tensile specimen, and the actual temperature alteration cannot be understood without information on the theory of experiment which describes the influence of the measurement system.

The physical approach represents the foundation of major trends in

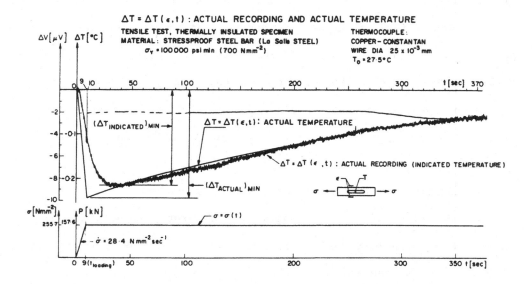

Figure 2 – Thermoelastic response of steel in tension: the indicated
         temperature of specimen (actual recordings), and the actual
         temperature (evaluated). Influence of measurement system.

contemporary experimental research. To apply it efficiently, it is nec-

essary to present a more comprehensive picture of the underlying con-

ceptions of physical reality.

2.4 Some Particular Issues

Of course, the meanings of relationships which have been evaluated

or derived using either the physical approach or the phenomenological

approach are often disparate. A discussion of results of the analytical

and the experimental research carried without information on the chosen

methodology usually yields results to which this known verse pertains:

> "Denn eben wo Begriffe fehlen   *
> Da stellt ein Wort zur rechten Zeit sich ein".

The fundamental difference between the results of the phenomenological
approach and of the physical approach when applied to real problems can be
illustrated by the following elementary example.

- It was customary, and it is still found fashionable by some young
engineers and researchers, to distinguish between theory and practice, and
to claim that the two are often contradictory. Such opinions are well
expressed in the following facetious German verse:

> "-Sag' Freund, was ist denn Theorie?
> - Wenn's gehen soll und geht doch nie!
> - Und Praxis? Frag' doch nicht so dumm.
> - Wenn's geht und keiner weiss warum!"**

Obviously, such a situation can only exist when - because of super-
ficial similarity rooted in a homonymous identity - a simple phenomenological theory is chosen too enthusiastically, to represent a complex
physical process.

Such a contradiction cannot exist between an empirical (experimental)
fact and a relevant physical theory that reflects a reality, consciously
and methodically simplified; such a theory cannot be mistakenly applied
outside of its range of validity, since all the simplifications (assumptions)
and constraints are clearly stated and justified.

However, it would be incorrect to underestimate the significance of

---

* "For if your meaning's threatened with stagnation,
   Then words come in, to save the situation".

** "Say, Friend, what is a 'theory'
   - When it should work but never does!
   And practice? Don't be so dumb.
   - When it works and no one knows why!"

a pure phenomenological approach in applied mechanics. Beside its intellectual values, the speculative approach - properly used within its limitations - gives answers to questions "what would happen if ..." which are necessary in analysis of real phenomena; it also produces results, many of which can be very helpful in understanding relations between major parameters of real processes.

One more problem, already mentioned, deserves attention: the growing significance of the short and long-term social factors in engineering. The most significant progress can be observed regarding the so-called "human engineering". This development started long ago. For instance, more than forty-five years ago, students taking a course in airplane design at the Warsaw Technical University have been advised that "an airplane is built around the pilot". One of the most rapidly growing fields of research in this direction is the bio-mechanics. Fortunately, since this is an almost non-controversial field where no phenomenological paradigms prevail and the modern technology requires reliable answers to many questions in order to progress, the rate of conceptual development in research is rapid.

A more general problem, the interaction technology-ecology-society, although extremely important, is approached more cautiously. Unfortunately, any new development in this field influences processes that are outside of pure intellectual considerations and outside of competence of engineering research laboratory or design office. One of the typical examples is, at present, the most progressive legislation regarding car safety, developed in the United States, and the related implementation problems.

These growing fields of engineering activity and the related new kinds of problems require an adequate development of reliable experimental methods and techniques.

Up to this point, the term model has been used in this text rather intuitively. According to experience, an intuitive understanding could be misleading, and it is necessary to be more precise.

The term "model" denotes usually all kinds of pictures of reality: the conceptual models in the analyst's mind; the picturesque models produced by human senses and mind; the symbolic and/or mathematical models; and the iconic models made of matter, fields and flowing energy, which are often denoted scaled, reduced or mechanical models.

However, in mechanics the term model is often used fairly loosely and the meaning of the same terms often is different or contradictory, depending on the unstated approach. For instance, the term "physical model" is being used to denote a conceptual physical model given by chosen sets of physical processes and related parameters, and also is being used to denote an iconic scale model. Very often the terms related to contradictory frames of reference, such as "law of mechanics" and "model" are used synonymously. The field is confused and requires a more explicit analysis.

3. MODELLING OF REALITY

It has been shown that the correct understanding of the methodology of development of models of real events is indispensable not only for a scientist but also for a modern engineer.

Usefulness of the developed or chosen model of a particular process can be assessed rationally only when such terms as reality, models of reality, reliability, and accuracy are sufficiently clearly defined. The chosen definition reflects of course our actual level of knowledge and of understanding of nature, and as such are subject to continuous refinements.

3.1   Outline of Approach.   Correspondence Principle.   Theory of Paradigms.

It is useful to define reality as a real universe of physical objects, fields and processes characterized by the related exchanges of energy[1,2]. This universe - of course - is not directly accessible, however, it is very useful to assume its existence.

Experience shows that it is safer and more rational to assume that the knowledge about this universe is collected according to the level of our understanding, our experience and, last but not least, to the accepted method of developing of new understanding, i.e., of developing new, more comprehensive models which allow us to describe and explain new empirical facts, Figure 3.

Consequently, science and technology develop a continuously changing images or models of real universe which are related to the perceived information.   This perceived universe consists of actually known and understood objects, fields and processes;   it usually represents a logically arranged system of models: hypotheses, axioms, conceptions, principles, theories, etc.   The perceived universe represents a very simplified and distorted but nevertheless very useful model of the real universe - the only one that can be used as an object of the experimental and speculative research.

From the point of view of experimental research in mechanics of solids and fluids, the term reality usually refers to the existence of bodies, fields, and systems, and to their responses to various inputs of energy. However, even in this narrow scope there is a sufficient space for ambiguity leading to formal and practical errors and misunderstandings.

In the last dozen or so years, we notice a slow but consistent diffusion of some basic ideas related to the aforementioned terms, and of the
resulting approaches, into the field of mechanics. It appears worthwhile
to mention two most characteristic contemporary approaches which result
in different interpretations of the development and meaning of our understanding of reality: the rational or intuitive application of the concept of correspondence based on the criterion of continuity of development of science[1-3], and the approach based on the concept of paradigm[4,5],
which denies the existence of the continuity in the development of understanding of reality.

The principle of correspondence and the derived concept of correspondence relation, developed by Niels Bohr in the period 1913-1922,
requires that a new more comprehensive theory contains the theory to be
replaced as a particular, usually limiting, case. The old theory being
replaced or generalized, represents at least formally, a limiting case of
the new theory when the new parameters attain extreme values, usually zero
or infinity - the new theory passes asymptotically into the old one when
new parameters can be neglected as variables. The new formulation must
be supported not only by new empirical evidence which could not otherwise
be explained but also by all the results supporting the old theory.

Being a methodological principle, the correspondence principle may
be interpreted as descriptive or normative. A descriptive interpretation
is not always possible or convincing. However, a normative interpretation,
when accepted as a criterion, is very useful and often indispensable in

such fields of applied science as applied mechanics, and thus, in experimental mechanics. Several examples of this indispensability will be given
subsequently.

The term "paradigm", formulated by Thomas S. Kuhn in the period 1959-
1970, denotes a theory or a set of theories, standards and methods developed to explain new empirical fact or facts not explained by the old theory. This set is not necessarily commensurate with the set to be replaced -
it only can supplement it or replace it completely. Consequently, no
"coherent direction of ontological development" can be observed in successive theories. Incommensurability between successive theories is a
rule rather than exception. Even the meanings of the same terms are usually different because they are related to disparate concepts. New concepts and related theories develop in the form of a scientific revolution.
No correspondence can be noticed among theories separated by a scientific
revolution, according to Kuhn. Kuhn's semi-tautological statement that "a
paradigm is what the members of a scientific community share, and, con-
versely, a scientific community consists of men who share a paradigm"
illustrates well the sociological roots of the notion of paradigm. This
statement, pertaining always to a certain period, describes one major
sense of the term paradigm. The other sense is related to the concrete
rule or theory called "puzzle-solutions" which can successfully replace
existing explicit rules or theories.

Using Kuhn's concepts and terminology, the statement that scientists
share a theory or set of theories should be replaced by a statement that
they share a paradigm or a set of paradigms, or - more precisely - that

they share a "disciplinary matrix" related to a particular discipline
and which is "composed of ordered elements of various sorts, each requiring
further specification".

It can be noted that the first approach is generally accepted in
physical sciences and that the modelling of physical reality is performed
accordingly[1,6-17]. However, one can note that in practice, particularly
in applied mechanics, it happens too often that several mutually incompatible models of the same reality are being used to describe the same
response of a body without any discussion of the assumptions made - a
clear example of a classical case of a spontaneous or intuitive paradigm!

In a more practical framework, the term paradigm is used to denote
a presently accepted theory of a particular event or process. Consequently,
one encounters often such terms as a useful paradigm, a comprehensive
paradigm, a historical paradigm, illogical paradigm, etc.

3.2 Some Side Effects of a Non-Critical Acceptance of Paradigms

In the Experimental Mechanics one very often notes a rather strange
habit of noncritical referrals to published engineering papers and books
as to a nonquestionable source of applied and basic knowledge and guidance.
It is often believed that it is not needed to refer to more fundamental
theories and concepts. Such an approach disregards the usefulness of
the correspondence principle, and the wisdom of the old proverb "errare
humanum est". This subject is usually very sensitive, therefore the examples presented below have been chosen from fields sufficiently remote
from Experimental Mechanics, as a kind of a prudent precaution.

- At the turn of the century an eminent physicist has published

results of his research regarding his discovery of a new form of electromagnetic radiation, which he named "N" rays. Subsequently, other physicists
published several dozen papers confirming the existence of these "rays".
Moreover, some authors claimed priority in discovering them. A few years
after this discovery another prominent physicist "annihilated" the "N
rays" - no more papers have been published.

- A few years ago, one of the most prestigious scientific journals
published a research note in which its author claimed that he had discovered "polywater", consisting of polymerized water molecules. Subsequently, several papers on this topic had been published, and some authors
claimed priority. A few years after this discovery, it has been shown that
the "polywater" is not a polywater but a particular aqueous solution of
a particular glass. No more references to "polywater" are being made.

- More recently, an engineering professional journal announced the
possibility of existence of 'perpetum mobile' engines!

A researcher committed to a phenomenological understanding of reality
is more likely to accept such attractive paradigms, than a researcher who
accepts the physical approach and the correspondence principle as the
major directing principles.

3.3   Typical Classes of Models of Physical Phenomena

Obviously, numerous - often quite different - models of the same
reality can be developed, and obviously none of them can represent all
the features of reality which is being modelled. Each model simulates
usually one particular feature of reality, that being the feature of
interest, or - infrequently - only a few of them. These features can be

selected purposely, on the basis of a general conceptual physical model,
or - what happens too often - on the basis of a phenomenological understanding of reality. This pertains to any model of reality: conceptual
physical models (conceptual pictures in the mind), mathematical models of
both types, physical and phenomenological, (symbolic pictures in the form
of graphs and equations), iconic scale models, or analog models.

Generally speaking, a model can be defined as a very simplified
picture of reality or - in a rather picturesque manner - as a caricature
of reality.

The major patterns of model development are presented schematically
in Figure 3. The meaning of terms used in this figure is explained in
the text.

The process of simplification can be performed fairly rigorously -
the end product, the main purpose of which is to understand the mechanism
of phenomenon, is usually called a conceptual physical model; such a
process is called a physical approach if the actual set of responses is
perceived, considered, selected and simplified, using the correspondence
principle.

Very often the selection of responses and the following simplifications are performed rather intuitively - such an approach, the main purpose of which is to describe rationally the observed processes, is called
phenomenological approach and the model developed is called a conceptual
phenomenological model.

As in the case of the basic theory of photoelasticity or ultrasonics,
the phenomenological approach is sometimes chosen consciously, because

CONCEPTION OF MODELS OF REAL SYSTEMS (BODIES AND PROCESSES)

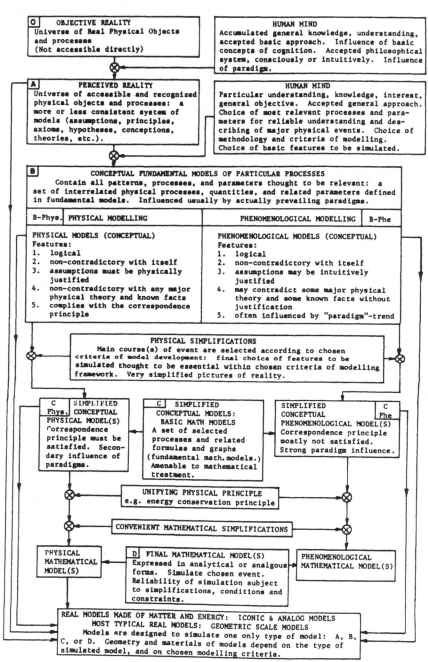

Figure 3 – Major patterns of model development: basic definitions and typical approaches.

of the lack of a sufficiently comprehensive understanding and description
of the actual physical process; however, it happens too often that such
an approach is chosen primarily for the sake of convenience.

Sometimes the difference between the physical and the phenomenological
approach is deeply rooted in fundamental philosophical beliefs. The
phenomenological approach in such cases is based on the assumption that
all the system parameters such as the modulus of elasticity, the dielectric
coefficient, etc., represent directly the inherent and permanent features
of the bodies, called properties. According to a more critical physical
approach, all we can determine are the responses of bodies to the various
forms of energy, and therefore the numerical values of particular responses must depend on the form of energy and its spacial and temporal
distribution, and on the chosen frame of reference (e.g., chosen mathematical models of basic phenomena).

Both approaches, physical and phenomenological, lead to a development
of particular models of reality, for instance, to designing of various
models of a given solid body or a given structure, or machine.

Of course, because the number of parameters representing any real
body or any real event is thought to be infinite, any model of reality
represents a particular simplification, according to the chosen criteria.
It is impossible to model any reality "as it is". It is necessary to
choose a criterion or criteria of modelling, i.e., to rationally select
an event or a set of events called here major physical event or events,
to develop a model the response of which will simulate the chosen physical
event(s), more or less rigorously. This fact imposes serious constraints

on the versatility of all analytical and experimental methods based on
particular models which are designed to simulate particular processes.

Depending on the chosen approach, the chosen criterion of modelling
and the chosen procedure of modelling, it is possible to distinguish
between the following classes of models, each of which is supposed to
represent adequately the process (response) of interest chosen as the
major physical event.

A. Physical Approach. Conceptual general physical model: a set of all
known physical processes and related physical parameters considered to be
significant; choice and selection of processes and parameters are performed accordingly to the level of knowledge and understanding of the
event being modelled.

- Simplified physical models: more or less rationally simplified
general physical model; simplification is performed according to the
chosen criterion of modelling.

- Basic physical-mathematical model(s) representing the final simplification of the chosen conceptual physical model.

- Final mathematical model developed as a result of simplification of
analytical relations which represent the basic mathematical model
(physical-mathematical model).

- Iconic models representing either one of the physical models or one
of the mathematical models.

- Analog models based on particular mathematical models, which describe
different physical phenomena. An analog model is a form of iconic
model, the basic physical event of which has been substituted by another

physical event; the course of this event can be presented by the same
equation; for example, the diffusion equation is used to model the transfer of heat, as well as the transfer of matter. This kind of relationship
is often called correspondence by analogy.

Because the physical-mathematical models usually simulate responses
and their mechanisms, they are often called structural models or system
models.

B. Phenomenological Approach. A simplified, or very simplified conceptual physical model, very often assumed intuitively, and usually based on
a very phenomenological model of reality.

- Mathematical model representing intuitive major simplifications of
the involved physical events, i.e., of the assumed physical model. The
phenomenological-mathematical models simulate usually the responses only;
their mechanism is not simulated. This is a classical "black box"
presentation.

It must be noted that the above-mentioned scheme is followed usually
in cases when a more or less rigorous analysis of the modelling process
is performed, i.e., when the reliability of the model to be developed is
analysed.

In everyday engineering practice, when the research or design activity is mostly based on routine procedures, standards or codes, the intuitively designed iconic models still are chosen almost routinely, and the
iconic (geometric) correspondence is commonly considered to be a sufficient
criterion of the model correctness. This practice often leads to unnecessary misinterpretations of experimental data since the iconic models

are - of course - models of models, limited to the simulation of selected responses only.

Formally, neglecting the rationale of the process of model development, the final form of a model of any real physical event can be:

- *iconic:* the response of a scale model is - or is thought to be - determined by the same physical parameters which describe the real physical event chosen as the major event of interest;

- *analog:* a physical event of interest is represented by an analogous physical event, the course of which can be presented - or is thought it can be - by the same equation(s), i.e., by formally the same mathematical model;

- *symbolic:* a major physical event is represented either by a set of analog relations between major physical parameters, or by a set of symbols representing responses of components of an equivalent system, or by equations representing relations between major parameters, i.e., by mathematical models.

Of course there is an inherent interrelation between all these classes of models. Iconic models are developed to simulate responses of particular physical systems, which in turn, are represented by selected physical parameters, usually assumed to be constant parameters of assumed mathematical models.

3.4   Major Simpliciations of Real Phenomena

For practical reasons, it is necessary and sometimes conceptually desirable to consciously simplify the models of the event of interest. Of course, it always is possible to assume the existence of a particular

hypothetical material characterized by a particular set of well-defined
relations between material parameters, and to investigate its responses
to a well-defined disturbance. As a next step, it is common to seek real
materials, the behaviour of which could be simulated by the assumed hypothetical materials; usually this is done intuitively, without any rational
scientific proof. That issue is outside of the scope of this discussion.

The process of simplification in development of models in mechanics
has been usually carried out using the so-called Ptolemean, or "black box"
approach. However, one can notice the growth of a new, physical approach,
typical for instance in materials science, the so-called Copernican or
structural (system) approach. The development of a new field of micromechanics can be taken as an example of a structural approach[18].

The most typical simplifications of real responses of bodies in
mechanics are:

- stochastic nature of actual events is replaced by a simple
deterministic one;

- inherently nonlinear character of all processes is replaced by
linear model(s);

- inherently interrelated physical events are replaced by sets of
independent relationships;

- discrete structure of materials is replaced by the concept of
a usually homogeneous material continuum;

- thermodynamic aspects of all deformation processes are neglected;
all strain states - static, dynamic, shock - are presented as isothermal
processes;

- inherently irreversible processes are modelled as reversible.

Such simplifications are often very useful and even necessary. However, it happens too often that the conclusions are drawn about real phenomena on the basis of predictions of their very simplified mathematical representations in such a manner that the limitations following from the assumptions are violated. The necessary - and sometimes unnecessary - simplifications are denoted by the term "idealization" which is apparently rooted in the belief that the ideal state is the simplest one. Well, one can argue, rather successfully, that the ideal state of nature is a most complicated one.

3.5   Comments on the Methodology of Simplication and Description: Atomistic and Continuous Schemes.

Two major methodological issues deserve particular attention: the choice of a most efficient methodology of simplification of reality and the choice of an atomistic (discrete) or continuous framework.

The manner of simplification is usually chosen within the limits determined by two extreme cases:

- a significant simplification of a physical model to such an extent that the resulting mathematical model, which represents a considerably approximate formulation of the problem, can be presented rigorously in its final analytical form;

- a less significant simplification of a physical model, which results in a more rigorous (realistic) mathematical model, final analytical formulation of which requires notable approximations.

In other words, the researcher usually chooses between a rigorous solution to an approximate formulation and an approximate solution to a more rigorous formulation.

The second issue - the choice of a discrete or a continuous framework for description - is even more significant than the first one. Unnecessary and confusing disparities between some popular theories and techniques used in Experimental Mechanics and the actually observed courses of events are rooted in a widespread belief that the concept of an isothermal homogeneous continuum can satisfactorily be used to describe all the phenomena of interest to Experimental Mechanics, e.g., such as interaction between radiation and deformed matter, photorheological processes, etc.

As it has been said above, it is methodologically correct, and practically very useful, to consider in Experimental Mechanics all bodies as complex systems, consisting of interacting particles. Consequently, the overall continuum framework can be developed as a result of a statistical treatment of inherently stochastic processes at the elementary level. In other words, a dualistic system of interacting particles and of a continuum is accepted[20]. Such an approach has been applied with great success in sciences, and in some fields of mechanics, particularly in micromechanics[18,21]. However, it is still not accepted in Experimental Mechanics, as a result of a belief that it is not needed for engineers. Pertinent engineering research publications are overlooked, e.g., Ref.[47], or papers published at the turn of the century referred to in Ref.[71]. Rejection of the dualistics system of interacting particles and of continuum unnecessarily delays progress in Experimental Mechanics.

3.6   Reliability and Accuracy of Models of Real Phenomena

The term "reliability" pertains to the actual correlation between
the flow of energy in the object and the model, i.e., to the character
of major physical events. "Accuracy" pertains to the numerical correlation between the values of signals processed by a measurement system.

"Reliability" of a model obviously depends on the chosen theory of
model development; it is therefore necessary to be quite specific regarding the major features of both major kinds of models, phenomenological
and physical.

In this framework, the traditional terms "precision" and "accuracy"
denoting the magnitude of scattering of results with respect to each
other and the nearness of the mean value of results to the so-called
"true value", respectively, represent parameters of the term "accuracy"
used in this text.

The term "accuracy" and the related term "precision" of measurements
are related to the answer to such questions as "how big is big" or "how
small is small". Obviously, the practicality, i.e., the safety factors
and economic factors, is of major importance regarding the engineering
choice of criteria for defining "big" and "small".

3.7   Property and Response

Rather unfortunately, the term "property" is being used simultaneously
in the rigorous and the colloquial sense. The rigorous meaning of the
term "property" which is rooted in the concept of a "true" mathematical
model conveys a strong philosophical message, being outside of the
analyzing capacity of experimental or analytical methods. Consequently,

the term "response" which is influenced by all the parameters of experiment, is the only appropriate term to describe rigorously the behaviour of bodies under observation[19] and therefore the meaning of the term "property" is a colloquial one when applied to real physical systems. Obviously, the term "property" is rigorous only when it denotes constant parameters of phenomenological mathematical models.

3.8 Summary of Common Features of Models of Real Phenomena

It is useful to summarize the already discussed common features of all kinds of models in terms of the assumptions, concepts, axioms, and the relevant principles.

Assumptions: (a) All physical processes are real; (b) all physical processes can be analyzed rationally: human mind is able to develop quite satisfactory models (reflections) of reality. Consequently, the models are invented and not discovered.

Concepts of: (a) Mass; (b) energy; (c) system, represented by responses which characterize the influence of a particular flow of energy; (d) information, collected by energy flow: no energy, no information.

Axioms: (a) Conservation of matter (substance and energy); and (b) the three laws of thermodynamics.

Principles: (a) Causality, simple deterministic and stochastically deterministic; and (b) correspondence and   correspondence relation.

Major Conclusions: (a) The so-called "laws of nature" are invented. (b) Models are invented according to the chosen, more or less rationally, criteria of modelling; consequently a model is able to simulate only certain features of reality, more or less reliably.

Typical Consequences: When applied to real physical events, the
principle of simple causality leads to various analytical expressions of
the relations between one state and the states neighbouring in time and
space. Two typical kinds of the analytical representation of the principle
of causality are used in modelling:

(a) very simplified, presented in the form of linear, ordinary
differential equations with constant coefficients:

$$\phi_1 \ (y, \frac{dy}{dx}, \ \ldots, \ \frac{d^n y}{dx^n}) \ = \ 0,$$

(b) more general, based on the concept of a system consisting of
interrelated events, and consequently presented in the form of nonlinear
partial differential equations with variable coefficients:

$$\phi \ (x,y, \frac{\partial y}{\partial x}, \ \ldots, \ \frac{\partial^n y}{\partial x^n}, \ \ldots [\frac{\partial^m y}{\partial x^m}]^k) \ = \ 0.$$

Consequently, the linear models are inherently, to a greater or
lesser degree, distorted representations of reality.

Very often a more specific form of presentation of the concept of
causality is used to emphasize the influence of the form of the response-
producing energy flow:

$$\sum_{k=0}^{n} a_k \frac{d^k s_r}{dt^k} = \sum_{\ell=0}^{m} b_\ell \frac{d^\ell s_e}{dt^\ell}$$

where s_r and s_e denote the response signal and the entrance signal,
respectively.

Another typical consequence of the physical approach is the recognition of the fact that the strain energy "accumulated" in an elastic body

during the process of reversible deformation is not equal to the
energy of the system, U; in other words the strain energy is not equal to
the work done by external loads, since

$$dU = dQ + \sigma_{ij}d\varepsilon_{ij},$$

where Q denotes the quantity of heat which must be exchanged reversibly
with the surroundings to maintain the body at constant temperature.

It must be recognized that the physical and phenomenological approaches in mechanics are intertwined[22,23]. However, a triumph of the
physical approach is best shown in the use of experimental results of
Eötvös in the formulation of Einstein's principle of equivalence on which
the general theory of relativity was based.

4. MODELLING IN MECHANICS

Discussion of modelling in mechanics must encompass at least the
concepts of models, systems, responses of materials and systems to the
input of energy, and the measures developed to characterize such
responses.

4.1 General Remarks

It has already been stated that the choice of a general frame of
reference must determine the structure of a model of a real phenomenon,
as well as the compatibility of the model to be developed with a more
general model, and the range of reliable prediction of the model to be
developed.

For instance, the choice of a coordinate system connected with the
earth centre, and rotating together with the earth about its axis, re-

quires application of the concept of the Coriolis acceleration to explain the patterns of motion of bodies on the surface of the earth. Such
a choice of a coordinate system leads logically to the concept of the
Ptolemaic planetary system, etc. The choice of a nonrotating coordinate
system connected with the earth eliminates the need for the application
of the concept of the Coriolis acceleration.

Obviously, questions such as "does Coriolis acceleration really
exist?" are meaningless if the chosen frame of reference of the invented
model is not specified.

The choice of a model is often influenced by cultural and psychological factors and in this respect Kuhn's concept of paradigm appears to
be very useful. Ptolemaic and Copernican models of planetary motion can
be made equivalent kinematically. However, Copernican model is (fortui-
tously) compatible with Newtonian dynamics, although Copernicus could not
have been aware of it, and the Ptolemaic model is not. The advantage of
the Ptolemaic model was its compatibility with the concept of ideal
curves, and with some elements of cultural life, especially with the generally recognized philosophical systems which yield particular sets of
conclusions accepted as major criteria for understanding of nature. The
advantage of Copernican model was its conceptual simplicity and the related
elegance. The disadvantage was its contradictions with the above-
mentioned criteria. Obviously, this contradiction could not be solved in
the frame of "true" models or "wrong" models. A strong influence of
paradigm is obvious.

Similar problems of modelling, rooted in culture, in education, and

in mentality, still exist. There is opposition to the teaching of
Darwin's theory of evolution among some groups of population because of
its incompatibility with certain criteria which are considered to be of
fundamental importance. Closer to home, among engineering population,
there exists a belief that any solution of an engineering problem must be
correct when presented in an analytical form, e.g., when represented by
a set of differential equations, because "mathematics can never be wrong".
A similar belief exists regarding the predictions of computer programs,
because "a computer cannot be wrong". Well, practical experience, sometimes very painful, recently led to formulation of a popular acronym "GIGO",
which inelegantly relates the reliability of the computer's predictions
to the quality of the assumed mathematical model:"garbage in, garbage out".

Regarding the patterns of propagation of radiant energy, we are concurrently using two mutually incompatible mathematical models: a cor-
puscular model and a wave model, each of which is able to explain different phenomena. As long as there has been a tendency to seek "the
truth", when one of both the models has been accepted as a "true discovery" the other was consequently rejected as a "false invention"[*].
Presently we are glad that only two simple models of radiation are sufficient for a description and explanation of a majority of engineering
problems, and we consider both models disparate but complementary – not
contradictory. Incidentally, in accordance with the correspondence

[*]Example of a paradigm may be found here, for as paraphrazing C. Maxwell:
"There have been two theories regarding the nature of light: particle
and wave. We believe in the latter because those who believed in the
former are dead".

principle, both of these models may be considered as particular cases of
a general model developed in the quantum theory.

Unfortunately, in mechanics it is still not customary to present and
discuss the features of mathematical models chosen as a foundation of
the theory to be presented or discussed.

Two known examples may illustrate this problem.

Example 1. Typical forms of mathematical models of physical processes are:

$$(F, \frac{dF}{dx}, a) = 0,$$

$$(F, \frac{\partial F}{\partial x}, \frac{\partial F}{\partial y}, \frac{\partial F}{\partial z}, \frac{\partial F}{\partial t}, a) = 0$$

The first equation represents a simple model of a relationship
between a physical quantity and its variation which is caused by an
infinitesimal alteration of one parameter; this is an approximation which
neglects the influence of all variables but one. The second equation
represents a simplified picture of a system consisting of interrelated
events.

Thus the ordinary and partial differential equations represent two
different approaches.

Example 2. Paradox of Zeno of Elea: "Motion is an illusion.
Motion does not exist, because at any given moment in time a body occupies
given points in space. Obviously there is no velocity at a point since a
point is dimensionless".

This paradox could have been solved only after a new mathematical
tool had been invented, based on the concept of   infinitesimal quanti-

ties and limit values. It took about two thousand years to solve this
problem.

A particular practical conclusion may be drawn: A comparison of
predictions yielded by one simplified mathematical model with predictions
yielded by another simplified mathematical model, in itself leads to
nowhere,[7,10].

4.2 Principles of Model Development

The main issues in the development of a sufficiently reliable model
are the determination or selection of major parameters which influence
the major course of events and a satisfactory selection of other ancillary
parameters. Of course, the process of determination, or selection, or
identification of major parameters occurs only within the framework of an
accepted set of sufficiently general mathematical models of basic phenomena. The process of simplication of real phenomena in mechanics, that
is the process of development of reliable physical and mathematical models
of mechanical events, is basically the same as that discussed in Section
3 and illustrated by Figure 3. It is presented schematically in Figure 4.

Each simplification - an analysed one, or an intuitive one - requires,
of course, formulation of the corresponding conditions and constraints.
The term "simplification" in Experimental Mechanics is related to a real
physical event, the picture of which must be compatible with the fundamental physical relations known to us; in other words compatible with
the pertinent, most general, paradigm actually available. For example,
an analytical relation attempting to describe the propagation of a
mechanical perturbance,   which lacks terms related to thermodynamic

MODEL DESIGNING IN ENGINEERING MECHANICS

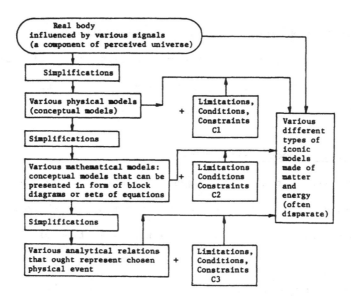

Figure 4 - Process of Model Designing in Mechanics

processes, is a simplified relation. Such a relation, in spite of this

limitation, is very often useful, but it can be dangerously misleading

if it is applied to more sophisticated cases.

Depending on the chosen criteria and the manner of simplification,

it is possible to design various different physical and mathematical

models of the same phenomenon, and various iconic ("real") models as well.

Material parameters which characterize any model in Mechanics are obviously

related to the chosen or the assumed fundamental mechanism of the pheno-

menon, i.e., to the chosen or assumed model.

The scheme presented in Figure 4 would imply that a model designing

process is a simple logical scientific process that may lead to different but noncontradictory simplifications. In reality the process of model designing is a complex fundamental task which requires a good understanding of the real physical process, clear formulations of the purpose of the model to be developed, and a good understanding of the physical significance of mathematical simplifications. Depending on the chosen criteria, the same model could be considered a sophisticated model or a crude model of reality. Unfortunately, such terms as reality, a problem to be solved, or more specific terms such as elastic, plastic, visco-elastic, are used often in two different senses: (i) in the physical sense, as more or less crude, but convenient and very useful, approximations of real responses or real bodies; (ii) in the phenomenological sense, as precisely defined quantities that exist in a hypothetical universe, that is in an assumed space governed by assumed physical laws.

A conceptual model developed according to the second approach cannot be accepted as a foundation of a rigorous solution of a problem; in this approach no proof is presented that the group Cl of limitations, conditions, and constraints, Figure 4, is satisfied and that the model actually represents the physical event which is being considered.

A rigorous solution for a model developed according to the first approach can usually be considered as a satisfactory solution of a problem, and can be reliably extrapolated within certain limits. Such a statement cannot be made regarding the phenomenological approach, without a sufficiently clear proof that the correlation between real and hypothetical events chosen as major events is satisfactory, both qualitatively and quantitatively.

Thus the starting point ought to be clearly stated: either an actual physical reality is to be modelled and the model development is rooted in the physical universe, or an assumed reality is to be modelled, with the developed model rooted in a hypothetical universe, which may or may not be commensurate with the physical universe.

There are merits and deficiencies in both approaches. In Experimental Mechanics, the physical approach is chosen, but very often - after analysis - solutions that were obtained in a phenomenological manner are utilized. The actual process of the first stage of model development in Experimental Mechanics is presented schematically in Figure 5. To complete the model design it is necessary to analyse the reliability and the accuracy of predictions of the intermediate stages of models. This problem will be discussed later.

Very often claims are made that a rigorous solution of a problem has been presented, or that a "true model" or "real model" has been developed.

Certainly, any model may be considered "real" or "true" in the same sense as the space of Riemann may be considered as real as the space of Euclid. However, from the point of view of Experimental Mechanics, models must be related to physical reality and consequently must be considered as approximations. According to Mark Kac[9]: "Models are, for the most parts, caricatures of reality, but if they are good, then like good caricatures, they portray, though perhaps in distorted manner, some of the features of the real world".

In applied sciences such as mechanics, the main role of models is to describe, to explain and to predict the behaviour of the real phenomena.

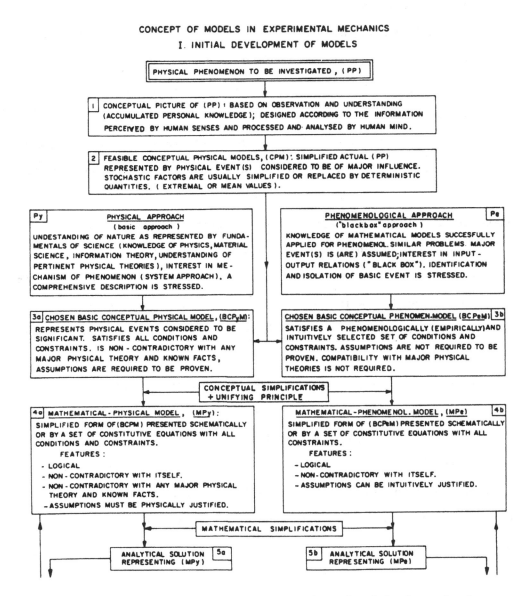

CONCEPT OF MODELS IN EXPERIMENTAL MECHANICS
I. INITIAL DEVELOPMENT OF MODELS

PHYSICAL PHENOMENON TO BE INVESTIGATED, (PP)

1 CONCEPTUAL PICTURE OF (PP) : BASED ON OBSERVATION AND UNDERSTANDING
(ACCUMULATED PERSONAL KNOWLEDGE); DESIGNED ACCORDING TO THE INFORMATION
PERCEIVED BY HUMAN SENSES AND PROCESSED AND ANALYSED BY HUMAN MIND.

2 FEASIBLE CONCEPTUAL PHYSICAL MODELS, (CPM): SIMPLIFIED ACTUAL (PP)
REPRESENTED BY PHYSICAL EVENT(S) CONSIDERED TO BE OF MAJOR INFLUENCE.
STOCHASTIC FACTORS ARE USUALLY SIMPLIFIED OR REPLACED BY DETERMINISTIC
QUANTITIES. (EXTREMAL OR MEAN VALUES).

Py   PHYSICAL APPROACH
              (basic approach)
UNDESTANDING OF NATURE AS REPRESENTED BY FUNDA-
MENTALS OF SCIENCE (KNOWLEDGE OF PHYSICS, MATERIAL
SCIENCE, INFORMATION THEORY, UNDERSTANDING OF
PERTINENT PHYSICAL THEORIES), INTEREST IN ME-
CHANISM OF PHENOMENON (SYSTEM APPROACH). A
COMPREHENSIVE DESCRIPTION IS STRESSED.

        PHENOMENOLOGICAL APPROACH   Pe
              ("blackbox" approach)
KNOWLEDGE OF MATHEMATICAL MODELS SUCCESFULLY
APPLIED FOR PHENOMENOL. SIMILAR PROBLEMS. MAJOR
EVENT(S) IS (ARE) ASSUMED; INTEREST IN INPUT -
OUTPUT RELATIONS ("BLACK BOX"). IDENTIFICATION
AND ISOLATION OF BASIC EVENT IS STRESSED.

3a CHOSEN BASIC CONCEPTUAL PHYSICAL MODEL, (BCPyM):
REPRESENTS PHYSICAL EVENTS CONSIDERED TO BE
SIGNIFICANT. SATISFIES ALL CONDITIONS AND
CONSTRAINTS. IS NON - CONTRADICTORY WITH ANY
MAJOR PHYSICAL THEORY AND KNOWN FACTS,
ASSUMPTIONS ARE REQUIRED TO BE PROVEN.

CHOSEN BASIC CONCEPTUAL PHENOMEN. MODEL (BCPeM) 3b
SATISFIES A PHENOMENOLOGICALLY (EMPIRICALLY) AND
INTUITIVELY SELECTED SET OF CONDITIONS AND
CONSTRAINTS. ASSUMPTIONS ARE NOT REQUIRED TO BE
PROVEN. COMPATIBILITY WITH MAJOR PHYSICAL
THEORIES IS NOT REQUIRED.

CONCEPTUAL SIMPLIFICATIONS
+ UNIFYING PRINCIPLE

4a MATHEMATICAL - PHYSICAL MODEL , (MPy):
SIMPLIFIED FORM OF (BCPM) PRESENTED SCHEMATICALLY
OR BY A SET OF CONSTITUTIVE EQUATIONS WITH ALL
CONDITIONS AND CONSTRAINTS.
      FEATURES :
- LOGICAL
- NON - CONTRADICTORY WITH ITSELF.
- NON - CONTRADICTORY WITH ANY MAJOR PHYSICAL
  THEORY AND KNOWN FACTS.
- ASSUMPTIONS MUST BE PHYSICALLY JUSTIFIED.

MATHEMATICAL - PHENOMENOL. MODEL , (MPe)   4b
SIMPLIFIED FORM OF (BCPeM) PRESENTED SCHEMATICALLY
OR BY A SET OF CONSTITUTIVE EQUATIONS WITH ALL
CONSTRAINTS.
      FEATURES :
- LOGICAL
- NON - CONTRADICTORY WITH ITSELF.
- ASSUMPTIONS CAN BE INTUITIVELY JUSTIFIED.

MATHEMATICAL SIMPLIFICATIONS

ANALYTICAL SOLUTION 5a
REPRESENTING (MPy)

5b ANALYTICAL SOLUTION
REPRESENTING (MPe)

Figure 5 – Basic approach to the construction of models in mechanics:
      first stage of development.

In contrast, in the pure sciences, the main role of models shifts towards

presenting controversial questions that generate progress by inducing polarization of opinions.

Since it is usually not possible to analyse or even to estimate the distortions of the phenomenon of interest caused by simplifications made during modelling, reliability and accuracy of a particular model predictions must be estimated experimentally.

The object of experimentation is either the transfer function itself (if known) which can be analysed using analog computer techniques, or the iconic, "real", models made of matter and energy. In fact, the inconic models may also be considered as a kind of analog computers.

According to the analysis summarised in Figures 3 and 4, one can build at least four different types of iconic models:- models which are not necessarily compatible with each other. The physical sense of the responses of different types of iconic models of the same phenomenon is not necessarily the same. Thus the conclusions drawn from the analysis of responses of "real" models regarding the reliability and accuracy of predictions of mathematical models ought to be drawn very cautiously. By the same token, a very simplified mathematical model may represent various physical phenomena, depending on the chosen sets of conditions and constraints, and on the physical meaning of all the coefficients. For instance, the ordinary linear differential equation with constant coefficients, such as the one below,

$$a_o s_o \ + \ a_1 \dot{s}_o \ = \ b_o s_i,$$

where s_i denotes an input signal and s_o denotes an output signal, may represent a response of a particular type of rheological body (Kelvin body), or a simplified response of a liquid-in-glass thermometer. The above equation is an analytical representation of a mathematical model that is applicable for description of several qualitatively different physical phenomena, if they can be simplified to a formally identical mathematical structure. The coefficients of such a model obviously may represent different physical parameters. Such formal identities of mathematical models and related analytical expressions are taken as the basis of various analog methods.

This was the first problem of Experimental Mechanics: "how to reliably describe reality".

The second problem is: "How to reliably observe reality". The second problem pertains to the reliability and the accuracy of data yielded by an iconic model, in relation to its own characteristic responses. The question is to what extent the applied loading and measuring systems influence the response, and how the influence of these systems on the experimental results could be minimized. It is not possible to analyse the degree of correlation between reality and the various models of it when the reliability and accuracy of collected information is estimated intuitively only. Consequently, it is necessary to analyse the patterns of response of loading and measuring systems which influence the process of producing, collecting, transforming and recording the desired information, to be able to evaluate the reliability and the accuracy of experimental data. To do that effectively, it is necessary to understand better the behaviour of real bodies.

4.3   Methods of Assessment of Model Reliability

It has been shown that a model, mathematical or iconic, can only simulate satisfactorily an event chosen as the object of the modelling criterion. In some cases several interrelated events are simulated. The reliability of a model refers to the degree of correlation between particular responses of a model and of the actual event being modelled, to the identical, in character, energy flows. Of course, only the so-called "indicated responses" can be observed and recorded, that is the responses indicated by measurement instruments.

As the responses of real physical bodies depend on all parameters of the supplied energy form and their variation in time, and since the obvious practical and economical reasons necessarily limit the scope of the response-evaluating experiments - 'thought' experiments or 'real' experiments - a particular experimental framework emerges consisting of: (i) standard signals, both deterministic and stochastic, specifying the time history of the supplied energy form; (ii) standard tests, specifying the geometric, physical and temporal parameters of the actual experiment; and (iii) the so-called calibration tests based on chosen analytical solutions, for instance, the solutions for typical strain and stress states.

Standard signals and standard tests are designed to represent the most typical or characteristic cases of real situations. Designing of experiments to assess the reliability and accuracy of predictions of the physical, mathematical and iconic models is one of the major problems of modern Experimental Mechanics.

5. GENERAL PATTERNS OF DETERMINATION OF RESPONSE OF REAL BODIES

5.1 Approach

The difference between a response to loads of a material per se and

of components or structures made of this material is fundamentally a
quantitative one only. Even a geometrically simple specimen can respond
to loads in a mechanically complicated manner. Consequently, the same
general approach can be chosen for modelling responses of materials and
of structures.

Let us denote by the term "response of material" any change of any
material behaviour (output signal) related to changes of external physical
parameters (input signal).

From the point of view of methods of Experimental Mechanics, the
following responses are of major interest.

Mechanical: described by the strain and stress tensors, their deriva-
                tives, and by the coefficients of mathematical models.

Magnetic: described by the magnetic permeability.

Electric: described by the dielectric tensor and resistivity.

Optical: described by the refraction tensor (related to the di-
                electric tensor), and by the derived quantities.

Thermodynamic: described by the temperature change in particular, and
                by the change of entropy in general.

It is known that all responses of real bodies to various input signals are coupled as shown in Figure 6. Some of these responses can be
detected by devices which drain directly the energy from the investigated
body and transform it into other forms of energy. Examples of such "active
transducers" are: accelerometers, force gages, mechanical and some
electric strain gages, thermocouples, etc. Some of the responses can be
characterized as changes in values of various material parameters, such

PATTERNS OF RESPONSE OF REAL BODY

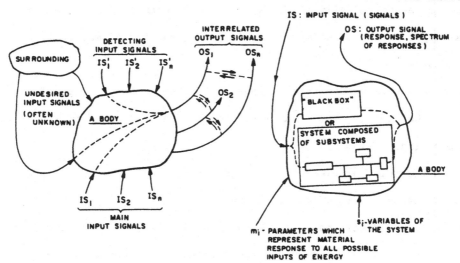

Figure 6 - Schematic presentation of a response to signals of a real body

    (a)   body in real surroundings,

    (b)   schemes of approach to analysis of response: phenomeno-
             logical ("black box") and physical (structural)

as specific heat, dielectric tensor, magnetic permeability, etc., or of

various derived parameters such as spectral index of refraction, spectral

dispersion, etc. Changes in values of such parameters require the detec-

tion of the modulation of an additional energy input introduced in a

suitable form, such as electromagnetic radiation, acoustic radiation,

electric field[19,24-30]; transducers based on such a concept are often

called parametric transducers.

    A scheme of a response of a real body is presented in Figure 6. The

main input signals consist of the desired input signal and of other (un-

desired) input signals coming from the surroundings. Very often it is necessary to produce an auxiliary (detecting) input signal to observe a particular kind of response of a body to the main signal.

Input signals, both main and detecting, are not, in general, completely independent parameters, Figure 6(a). They depend on the relation between the ability of a body to absorb the supplied energy and the ability of a source of the input signal to supply energy.

Since in principle all the input signals influence all responses of a body, all output signals depend on all input signals,[8]:

$$OS_n = OS_n(IS_1, IS_2, \ldots, IS_k, \ldots, \delta IS_1, \delta IS_2, \ldots, \delta IS_k, \ldots),$$

where OS_n denotes the nth output signal, IS_k denotes the kth input signal, and δIS_k denotes the variation of the kth input signal.

All responses represented by output signals are interrelated (coupled), Figure 6(a),

$$OS_1 = OS_1(OS_2, OS_3, \ldots).$$

Obviously, interactions between a body and its surroundings can be represented by an infinite number of input signals. Such interactions can be adequately analysed and described only when the body is considered a part of a more general system (in the thermodynamic sense).

It is convenient to present responses of real bodies in terms of the materials parameters "m", and the variables of the system "s",[8]. Material parameters m_i, represent material responses to all possible inputs of energy.

A body under observation can be considered to be a system which, in turn, may be described in various ways. Usually a choice is made between:

(i) a phenomenological presentation = "black box" presentation, and

(ii) a structural presentation = physical presentation = system presentation.

For instance, the relation between the refractive indices in the visible band of electromagnetic radiation and the structure and state of material can be presented as: (a) an intrinsic property of material, i.e., a "constant", and (b) can be related to the fact that the refractive index for visible light is determined essentially by electronic polarizability. The greater the ease with which the electron cloud can be distorted, the greater the polarizability, the greater the refractive index. Depending on the material parameters, the refractive index may be represented by a real or a complex number (tensorially).

In the first presentation, no condition is given for the dependence of the velocity of radiation on the spectral frequency, i.e., on the wavelength of radiation. This information must be added to the constitutive equation artificially.

The second presentation directly implies that the interaction between electromagnetic radiation and matter is a dynamic process, and therefore must depend on the excitation frequency.

As another example, regarding the mechanical and optical creep effect, the nonlinear relation between strain and birefringence for polystyrene can be presented as a relation: (i) resulting from the elastic and plastic components of polystyrene: phenomenological approach, accepted, e.g., in photoelasticity; or (ii) resulting from the orientation of anisotropic styrene molecules: system approach, typical, e.g., in materials science and modern material engineering, and infrequently applied in engineering photoelasticity. The same remarks as for the former example apply also for this case.

5.2   General Presentation of Body Responses

In the phenomenological approach, the response of a body given in Figure 6(b), as represented by relation between the output signal and the input signal, can be presented in the form,[8]:

$$r = \frac{OS}{IS} = r(m_1, m_2, \ldots, m_i, \ldots; \; s_1, s_2, \ldots, s_j, \ldots).$$

Since in general the input signal depends on time, so also do the output signals.

For practical reasons, it is necessary to restrict the unlimited number of material and system parameters to a sufficiently small number, representing the major course of events, the major features of the system, and the major related responses. In the "black box" approach the response of a body can be presented in a form:

$$r' = \frac{OS}{IS} = r'(m_1, m_2, \ldots, m_n; \; s_1, s_2, \ldots, s_k).$$

The physical or system approach in studying materials behaviour is based on the recognition of the fact that the concept of continuum loses its usefulness at the molecular level, and that strong interactions exist between certain events. Consequently many "black boxes" should be represented by corresponding sets of subsystems.

When material parameters and variables of a system are restricted to a finite number, in the system approach the response of a body can be presented as follows:

$$R' = \frac{OS}{IS} = R'(r_1', r_2', \ldots, r_\ell'),$$

where the functions r_i' which are defined according to the preceeding
equation represent the influence of particular subsystems, restricted to
major parameters.

A body can be thought of as a device which modulates, transforms
and transmits the flow of energy, regardless of the origin: the desirable,
the detecting, or the undesirable external energy flow.

The modulation and/or the transformation of energy flowing through
a body can be conveniently described in terms of a transfer function,
Figure 7, which can be represented as an operation on the input signal:

$$s_o(t) = G[s_i(t)].$$

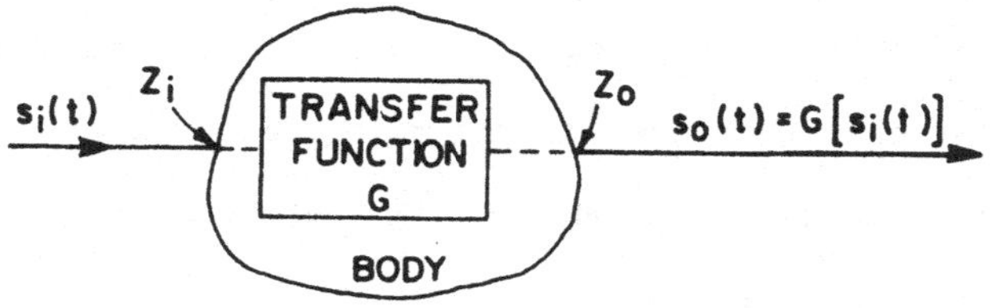

Figure 7 - A body as an energy transforming and transmitting device.

The ability of a body to absorb (to drain) energy and to allow the
drain of energy, can be conveniently described by means of the concept
of impedances, Z: input impedance Z_i; output impedance Z_o.

Obviously, impedance is related to a particular form of energy;
consequently, a body can be characterized by several impedances each one
related to a particular form and parameter of energy.

In general, the transfer function G, can be defined as a mathematical
operation which transforms an input signal into an output signal. This
function may be evaluated from a mathematical model of a body (constitutive equation), or it may be evaluated experimentally for one or more

forms of input signals and presented in form of an empirical function
or in an analog form as a diagram.

When a physical body is considered as an energy transforming and
transmitting device, it is necessary to accept that the manner of such
transformation and transmission ought to depend on all the parameters
which characterize the chosen forms of energy. Consequently, any information and any pattern of response of a body or any prediction of body
response is incomplete and often of little theoretical or practical value,
unless the pertinent parameters of the corresponding energy form are
given. Some examples are given in the table below.

Table 1 - Parameters of Response and the Related Information on
          Parameters of Energy Form which Produces or Detects a
          Response

Parameter of Response	Pertinent Parameter of Energy Form
Dielectric Coefficient	Frequency of Electric Field
Birefringence	Wavelength of Radiation
Yield Stress	Rate of Loading (Rate of Strain)
Curvature of Light-Path in Stressed Bodies	Wavelength of Radiation
Thermoelastic Temperature Alteration	Rate of Strain Energy

The main issue in developing a sufficiently reliable model is the
choice (sometimes called the "identification" in a phenomenological
language) of the primary parameters influencing the major course of the
events under consideration and a satisfactory selection of the secondary
parameters.

Since any information on response of material bodies to any input signals is produced by means of suitable loading and measuring systems which influence the process under observation, the major patterns of response of all coupled signals are of paramount interest,[7].

Typical signal interactions in a real body (system, subsystem) are presented schematically in Figure 8. Energy and information path in a typical testing system is illustrated by Figure 9.

PATHS OF SIGNALS AND INTERACTIONS IN REAL BODY

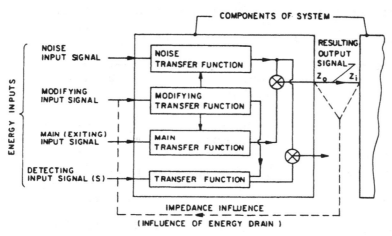

Figure 8 - Path of signals and interaction of signals in real body.

A sample of various interpretations of the same phenomenon depending on chosen mathematical model is given below.

Example 1. Two sets of incompatible conclusions about behaviour of low-carbon steel may be, and often are, drawn from the same experimental data, Figure 10, depending on the chosen mathematical model.

According to the common, phenomenological mathematical model of the experiment, the so-called upper yield stress $(\sigma_Y)_u$, and lower yield stress $(\sigma_Y)_\ell$ are material constants.

According to the less simplified physical-mathematical model of the

ENERGY AND INFORMATION PATH IN EXPERIMENTAL SET-UP

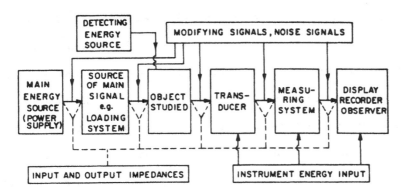

Figure 9 - Energy and information path in a testing system.

MATHEMATICAL MODELS OF TENSILE TEST

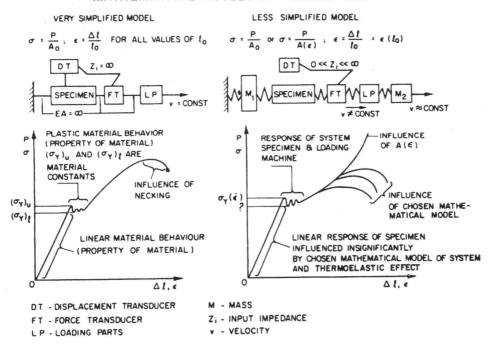

Figure 10 - Example of interpretation of experimental results according to chosen mathematical model. Object: response of specimen made of low-carbon steel to axial load produced by a testing machine (stress-strain curve).

experiment, the "upper yield stress" depends on the strain rate, and the "lower yield stress" depends on the dynamic characteristics of the system specimen-loading machine; since the so-called lower yield point represents only a parameter of the dynamic response of the system specimen-testing machine, it therefore cannot be considered as a material constant.

In addition, the slope of the semi-linear portion of the stress-strain curve depends on the thermodynamic conditions of the tensile test. For a positive coefficient of thermal expansion, the slope of this curve will be smaller for isothermal test conditions (low strain rate) than for adiabatic test conditions (high strain rate). The corresponding material parameters are often referred to as the "static modulus of elasticity" and the "dynamic modulus of elasticity", respectively. In principle then, the value of Young's modulus depends on the rate of change of the stress field. However, Young's modulus is a constant, when it is defined as a parameter of hypothetical Hooke's body.

Consequently, it is clear that the values of all the so-called material constants are related to the parameters of experiments. This fact has been recognized long ago by experimental scientists and engineers, and the conditions for determination of comparable results needed in engineering practice have been explicitly formulated in the various standards for testing of materials.

It is interesting to note that the lower yield point was used to verify the correctness of some phenomenological theories.

Example 2. There exists a particular coupling between the deformation of an elastic body and the alteration of its thermodynamic quantities

that is manifested by perturbations of the temperature field of a body
subjected to deformation at adiabatic conditions. It is customary to
call this coupling the "thermoelastic effect".

The phenomenological theory of this effect based on selected assumptions - one of which is the restriction to small deviations from thermodynamic equilibrium[31] - yields an analytical relation between the pertur-
bation in value of absolute temperature T, (for any given thermal
expansion coefficient α_T, specific heat at constant strain c_ε, material
coefficients E and ν), and the values of strain and stress components, ε_i
and σ_i:

$$\frac{dT}{T} = -\frac{\alpha_T}{c_\varepsilon} d(\sigma_{kk}) = -\frac{\alpha_T}{c_\varepsilon} \frac{E}{1-2\nu} d(\varepsilon_{kk}).$$

Temperature alteration ΔT, can thus be approximated by

$$\Delta T = -T\frac{\alpha_T}{c_\varepsilon} \Delta\sigma_{kk} = -T\frac{\alpha_T}{c_\varepsilon} \frac{E}{1-2\nu}\Delta\varepsilon_{kk},$$

where c_ε is practically equal to specific heat at constant stress c_σ,
for many engineering materials.

The above relation yields satisfactory predictions, when the rate
of stress is sufficiently high, i.e., when the actual thermodynamic process is close to being adiabatic.

Results of an actual experiment are presented in Figure 2. The
scattering of the recorded data is relatively very low, that is, the
accuracy of measurement represented by the indicated temperature graph
is high. However, the immediate apparent reliability of measurements is
rather low during the first twenty seconds, as indicated by serious discrepancy between the actual temperature graph (solid line) and the

indicated temperature graph. Obviously, it is necessary to design a
reliable mathematical model of heat transfer between the object and the
temperature sensor to utilize properly the accurate data supplied by the
measurement system. In such a frame of reference, the classical term
"accuracy of measurement" cannot be understood as an offset of measurement
instruments any more, and is not directly applicable. The time-dependent
discrepancy between the actual and indicated temperature can be understood, explained, described and predicted only on the basis of a commensurate mathematical model.

One can note that in a pure phenomenological approach, it is often
assumed that the functions representing the performed experiment only, in
fact. represent adequately the response of the body under investigation.
However, it is shown that only the knowledge of the theory of the performed experiment makes it possible to correctly interpret, to interpolate,
and to extrapolate the experimental results, in order to evaluate the
actual body responses.

5.3   Levels of Experiments

One more problem needs clarification, namely the distinction between
experimental investigation, industrial testing, and pilot testing.

It follows from the foregoing analysis that experimental investigation must be based on fairly rigorously developed physical and the resulting mathematical models of the process under study and on the defined
sets of conditions and constraints. The aim of experimental investigations
is usually to study the phenomenon of interest in a more general scope to
establish unknown relationships, to study the reliability of analytical

solutions, or to establish a scientific basis for testing. Experiments

which do not satisfy such conditions should be classified as testing or

pilot testing.

Industrial testing is based on sets of standards (codes) which, in

turn, are founded on the results of analytical and experimental investi-

gations. Specifically, testing is based on well defined physical and

mathematical models of phenomena, on well defined methods and techniques

of investigation, and on well defined sets of conditions and constraints

possibly including influence of surroundings. The values of all major

parameters of experimental techniques, of the conditions and constraints,

are specified numerically subject to given tolerances (e.g., see D.I.N.

Standard for evaluation of Poisson ratio). As a result, testing yields

results which are reproducible, comparable, analysable, and which can

usually be applied to check the reliability of new theories. Experiments

which do not satisfy such conditions may be considered pilot tests only.

In pilot tests, the same techniques are usually applied as in experi-

mental investigation or in industrial testing; however, pilot tests are

not based on pertinent physical and mathematical models, are not con-

strained by a corresponding set of conditions, and are not prescribed by

testing standards. Conditions of pilot tests are usually chosen intuitively

only, according to the aim of testing, to the financial constraints and often

according to the personal preference of the experimenter. The aim of

pilot tests is usually to obtain information on the general patterns of

a phenomenon, quickly and inexpensively. Consequently, the results of

such tests are related to the particular test and the particular experi-

menter; they are hardly comparable with the results of other tests,
since theoretical bases for such comparisons do not exist.

Unfortunately, it happens too often that the results of such pilot
tests are mistaken for the results of industrial testing or of experimental investigations.

6.   ELEMENTS OF INFORMATION PRODUCING PROCESS

6.1   Information

It has been shown by several authors[1] that the flow of information
requires a related flow of energy, and consequently it influences the
entropy of a closed, isolated system: "flow of information ≡ flow of
energy", that is "no energy → no information".

Energy flow in any closed and isolated system is related to the
change of entropy S, that is constrained by

$$\Delta S \geqslant \frac{\Delta Q}{T}; \quad S_B - S_A \geqslant \int_A^B \frac{dQ}{T} ,$$

where positive ΔQ denotes the addition of heat to a subsystem and T
denotes absolute temperature, and the equality is reserved for hypothetical reversible processes.

In a real closed and isolated system the total amount of energy remains constant but the total entropy has a tendency to increase; that is
the flow of energy from one subsystem to another results in a degradation
of energy.   This may be shown by studying the exchange of ΔQ amount of
heat between two subsystems, X and Y.   If we assume that energy is transformed from X to Y,

$$\Delta S = \Delta S_X + \Delta S_Y \geqslant - \frac{\Delta Q}{T_X} + \frac{\Delta Q}{T_Y} .$$

Since for transport of heat $T_X > T_Y$, thus $\Delta S > 0$.

The concept of negative entropy has been introduced by Schrödinger; L. Brillouin has named it "negentropy" N, such that $N = - S$.

Increase of information represents a negative contribution to entropy, and is equivalent to an increase of negentropy. As a result, it is not possible to perform any observation without increasing the entropy of the system. During each experiment the negentropy is transformed into information.

As an interesting aside it may be noted that the transformation and processing of information by the human mind is, as yet, outside of the realm of thermodynamic considerations.

To obtain information about any state or process in any object, it is necessary to drain energy from a corresponding region of this object, to transform this energy in a suitable manner, and to supply the transformed energy to a display instrument or a recording instrument.

Very often, when inducing the desired process in an object by means of a suitable device, i.e., power supply that supplies a particular form of energy to the object in a particular manner, it is necessary to supply an additional energy flow, called detecting energy, to detect the alteration of chosen parameters of process of interest. A set of components needed to produce the necessary energy, to influence the flow of energy through the object, to transform one form of energy into another form, and to transform the final form of energy into displayed or recorded data, is called measurement system.

The flow of energy is a major physical event that characterizes

performances of the testing, measuring and recording systems. A particular form of energy supplied to a body or drained from a body is called
an "input signal" or "output signal", respectively. The input and output
signals are characterized by temporal and spatial distribution of the
energy flow. The output signal carries the information of interest which
depends, in general, on all the parameters describing the energy form of
an input signal.

Recognition of the fact that the collecting of information requires
expenditure of energy made it possible for Brillouin to solve in 1951 the
thermodynamical paradox of Maxwell's demon, invented in 1871.

A demon of minute size was assumed to operate a small trap-door
between two vessels filled with gas at uniform temperature, Figure 11.

By permitting the fast molecules to enter vessel B and the slow
molecules to enter vessel A, the demon ought to be able to create a difference in temperature and pressure between the two vessels.

Brillouin has pointed out that if the demon were to identify the
faster molecules and the slower ones, he should have to illuminate them
in some special way, thereby supplying energy to the system and causing
an increase in entropy that would more than compensate for any decrease
in entropy produced by separation of molecules. This solution of Maxwell's
paradox complies with the second law of thermodynamics.

It is necessary to be more specific regarding the patterns of energy
flow through a system. Flow of energy through a system is modulated by
the process of interest; that is, some particular parameters of a given
state of energy are altered. That alteration usually depends on all the

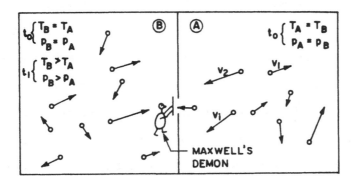

Figure 11 - Maxwell's demon as a violator of the second law of thermodynamics. Invented in 1871 by James Clerk Maxwell,
annihilated in 1951 by Leon Brillouin.

parameters that characterize the chosen form of energy. A sample of such

parameters associated with particular forms of energy is: (1) electric

energy: form of signal, frequency or repetition rate, amplitude, phase

angle; (2) acoustic energy: spectral, spatial and temporal power distri-

bution, dominant frequency, total power; (3) radiant energy: spectral,

spatial and temporal power distribution, dominant spectral frequency

(wavelength), total power, coherence length, degree and state of polari-

zation.

The lack of a complete set of information on the state of the

applied energy may lead to misunderstandings in the interpretation of the

obtained results.

Of course, the term "state of energy" denotes the state of energy

accepted by the object.

6.2 Measurement

Measurement may be defined as a process of comparing numerical value

of a physical quantity with numerical value of the reference physical quantity of the same kind, which is taken as a unit value.

Since the process of measurement - being a process of exchange of energy between an object and a measuring system - always influences the quantity being measured, the specification of the unit quantity, such as the numerical values of density, strength, etc., is meaningless unless the corresponding principle, technique and the conditions of measurement are specified. For the same reason any information on the accuracy of measurements is valueless, unless the technique of measurements and the reference values are specified.

Obviously, "true values" do not exist - they are theoretically impossible. Value of any measured parameter is influenced by the process of measurement, the parameter itself being dependent on the chosen mathematical model. Some examples are: dimensions of a body, temperature of a body, modulus of elasticity, Poisson ratio, transition temperature, etc.

A physical quantity is represented by a number and by a dimension, where the dimension may depend on the chosen mathematical model.

It is customary to distinguish between direct measurements and indirect measurements, although the difference is not always clear. A typical indirect measurement is always based on a set of mathematical models which describe relations between the quantities of interest. For instance, one can never measure the stress or strain. One can measure only the amplitude and the form of an incoming signal which is related to the quantity of interest by a set of chosen mathematical models — see Figure 12.

SCHEME OF STRESS MEASUREMENTS

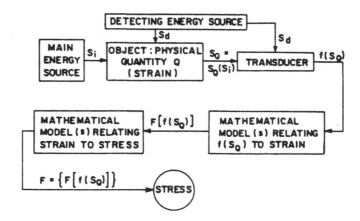

Figure 12 – Schematic presentation of strain measurement

In practical applications, one or two energy flows through a body under investigation are produced, as outlined below: (a) energy is supplied to an object by the main signal, part of which is drained from this object to carry the desired information; (b) supplementary flow of energy through the body is produced, which carried the desired information, when the main signal does not supply directly the needed information-carrying energy.

Energy is drained from an object by means of particular energy transforming devices such as gages, transducers, etc. The transducers can be: (1) active transducers (self-generating); and (2) passive transducers (require a supplementary energy flow).

The exchange of energy between an object and a transducer is usually not limited to the exchange of the energy form of the main signal. The interactions between a transducer and an object, and the manner of transformation of the drained energy into the transducer output signal are also of major importance.

It is convenient to characterize a transducer by means of:

(a) a transfer function and the related responses to chosen input signals;

(b) input and output impedances related to the chosen form of energy; and

(c) impedances related to other forms of energy that may be exchanged between the object and the transducer.

6.3 Systems

One uses the concept of "system" to denote a set of components that produce, conduct, modulate and transform energy from one form into another. A matrix of elements is not a system if no energy flow exists between the elements. The energy flow between two components depends on both components, that is, it depends on the major parameters of response of both components.

Consequently, a system of coupled components is essentially different from a chain of components. In a chain of components the level and the form of an input signal supplied to a certain component must not be influenced by the response of this component. In such a case, one refers to a "measuring chain" of connected elements. In a system consisting of several components, however, the energy flow between two components may depend on both components. Here the concept of a "measuring system" is required.

Since any information on the response of material bodies to input signals is produced by means of a suitable loading and measuring system, and since the resulting system dynamics influence the process under obser-

vation, the primary patterns of response of all coupled systems components
are of paramount interest.

Typical signal interactions in a real system, subsystem, or component,
are the same as those in a real body, which are schematically presented
in Figure 8. The output signal depends not only on the main input signal
and on the main transfer function; it depends on the modifying signals
as well, and is always influenced by noise, which may be produced extern-
ally and/or by the instrument itself.

Depending on relations between the output impedance of the signal
transmitting component and the input impedance of the signal receiver,
the modifying input signal may or may not be significantly influenced.
The energy and information path in a typical loading and measuring system
is illustrated by Figure 9.

Each component of a loading and measuring system may be represented
in terms of a system transfer function, or in terms of system impedances.
The body, responses of which are observed together with the responses of
loading and measuring systems, represent a subsystem, the responses of
which may or may not be linear.

Relations between the chosen parameters of response of a body and
the chosen parameters of the form of energy of the output signal may often
be presented as linear relations when the flow of energy through a body
is kept within certain limits; that is when other parameters which
influence the amplitude and the form of the output signal are kept within
certain limits. For example, relation between the output voltage and the
resistance of a gage in Wheatstone bridge connection may be accepted as

linear only within a range that, to a large degree, depends on the input
impedance of the voltmeter. The same pertains to the response of the
overall system, which is influenced by all the system components.

Because of many practical advantages, linear systems are, in general,
the first choice in experimental research.

6.4   Standard Forms of Signals

Responses of bodies in general and of measurement instruments in
particular, as well as functions representing responses of mathematical
models of bodies, depend on the shape of the input signal.

Since the results of measurement depend significantly on the responses of measuring instruments to signals produced by the object investigated, the knowledge of the character of responses to various shapes of
signals is of a paramount interest. It is, however, not possible and also
not necessary to analyse and to describe the responses of investigated
objects to all the possible forms of input signals. Consequently, it is
not necessary to know the responses of gages, transducers and measurements
instruments to all possible forms of signals. It is sufficient to know
the responses to certain standard signals in order to evaluate the performance of instruments and to choose in a rational manner the instrument
most suitable to given conditions.

Several standard signals, both of a deterministic and stochastic
character, are used. The most important of deterministic standard
signals include:

(1) Unit step signal (Heaviside):

$$s(t) = s_o <t - t_o>^0 = s_o H(t - t_o)$$

(2) Limited unit step signal:

$$s(t) = s_o[<t - t_o>^0 - <t - t_1>^0].$$

(3) Ramp signal:

$$s(t) = \dot{s}_o <t - t_o>^1.$$

(4) Limited ramp signal:

$$s(t) = \dot{s}_o[<t - t_o>^1 - <t - t_1>^1].$$

(5) Unit impulse signal (Dirac):

$$s(t) = Au_1(t),$$

where:

$$u_1(t) = \frac{1}{A} \lim_{T \to 0} (\frac{A}{T}), \quad A = \text{const.}, \text{ and } \int_o^T u_1(t)dt = 1.$$

(6) Stationary periodic signal:

$$s(t) = K \sin\omega t, \quad \text{where } K, \omega = \text{const.}$$

Parameters describing the performances of modern instruments are always related to the above standard signals. Obviously, the concept of "accuracy per se" of an instrument is meaningless. It is also clear that the concept of accuracy of an instrument must be related to the form of input signal and must encompass two kinds of errors of measurements: transient and steady state.

6.5 Impedance.

Generalized impedance is a concept that has proven very useful in experimental research in mechanics.

Impedance Z, is a convenient measure of the ability of a body or a system to transfer energy or power. Given a flow of energy between two bodies A and B, one distinguishes between: (a) input impedance Z_i which characterizes the ability of body B, say, to accept (drain) energy (power); and (b) output impedance Z_o which characterizes the ability of body A, say, to supply energy (power).

The flow of energy from body A to body B depends both on $(Z_o)_A$ and $(Z_i)_B$. Impedance can be defined for any form of energy, i.e., electrical thermal, mechanical, acoustic, bioelectrical, etc.

The generalized impedance Z_g, may be presented in a generalized form as a ratio of two variables, the product of which has the dimension of power P, for instance an effort variable q_1, and a flow variable q_2:

$$Z_g = \frac{q_1}{q_2}, \quad \text{or} \quad q_2 = \frac{q_1}{Z_g}, \quad \text{where } q_1 q_2 = P.$$

In the case of input impedance, the product presented by the above relation represents an instantaneous rate of energy withdrawal from the preceding body (source of energy). Impedance is, of course, a dynamic quantity.

The variables q_2 and q_1 are commonly denoted as follows: (a) flow variable ≡ through variable ≡ extensive variable ≡ kinematic variable ≡ global variable; (b) effort variable ≡ across variable ≡ intensive variable ≡ local variable.

According to the last of the above relations the power drained by the $(n)^{th}$-component described by the generalized input impedance Z_{gi} from the $(n-1)^{th}$-component being a source of the signal, is given by:

$$P = q_{i1} q_{i2} = q_{i1} \frac{q_{i1}}{Z_{gi}} = \frac{(q_{i1})^2}{Z_{gi}} .$$

Consequently, to minimize the power drain in a system, it is necessary to increase the input impedance.

The condition for a low demand of energy flowing from the $(n-1)^{th}$ to the $(n)^{th}$-component is:

$$\frac{(Z_o)_{n-1}}{(Z_i)_n} << 1.$$

The concept of mechanical impedance (developed about fifty years ago) is related to the concept of mechanical mobility, (developed about forty years ago). During the last ten years these concepts have been widely used in engineering research and development.

The concept of impedance is especially useful in experimental determination of the dynamic response of mechanical systems in a wide range of frequencies. Because of it, the mechanical impedance investigations represent a rapidly growing field of research in Experimental Mechanics.

Three main directions of such a research are currently represented:

(a) experimental determination of dynamic response of mechanical components and systems which are too complicated to be amenable to an analytical investigation;

(b) prediction of the overall performance of a mechanical system in a wide frequency band of loading forces on the basis of dynamic responses of system components;

(c) estimation of loads acting at certain points in a mechanical system from the measurement of the displacement time derivatives and vice-versa.

In mechanical systems, the force is usually the quantity to be measured; it may therefore be taken as the effort variable (local variable). The second variable, the flow variable (global variable) is defined as

$$\text{global variable} = \frac{\text{power}}{\text{force}} = \text{velocity}.$$

Hence, the mechanical impedance:

$$Z = \frac{\text{force}}{\text{velocity}} = \frac{F}{V},$$

and the mechanical mobility:

$$M = \frac{\text{velocity}}{\text{force}} = \frac{V}{F}.$$

It is customary to define: (a) point impedance; force and velocity are measured at the point where the force is applied; and (b) transfer impedance: velocity is measured at any arbitrary point in a mechanical system.

When the loads do not alter with time, the power absorption will be zero after a certain amount of energy has been supplied to the system. In such cases, it is more convenient to introduce different concepts and to relate them to the energy stored E.

These concepts are: (a) generalized static mechanical stiffness, S_V:

$$S_V = \frac{\text{force}}{\text{displacement}} = \frac{F}{\ell},$$

and (b) generalized static mechanical compliance:

$$D = \frac{\text{displacement}}{\text{force}} = \frac{\ell}{F}.$$

Since:

$$E = F\ell = F \int V dt,$$

hence

$$s_V = \frac{\text{(local variable)}}{\int \text{(global variable)} dt} \quad .$$

Measurements of mechanical impedance are closely related to measurements of the general frequency response. The response of a structure to a sinusoidal input signal is described by two relations: ratios of maximum values, $|F/V|(\omega)$, and phase angles, $\psi = \psi(\omega)$.

The theory of mechanical impedance is still developing. It should be noted that the vectors F and V depend on time and position and that mechanical impedance is actually a tensorial quantity.

6.6 Transfer Function

Mathematical models of dynamic response of linear systems are often represented by ordinary linear differential equations with constant coefficients, i.e.,

$$a_n \overset{n}{s_o} + \ldots + a_1 \dot{s}_o + a_o s_o = b_m \overset{m}{s_i} + \ldots + b_1 \dot{s}_i + b_o s_i, \qquad \text{(A)}$$

where: s_o, s_i are the output and input signals, respectively; a, b are constants which represent physical parameters; and $s^n \equiv \dfrac{d^n s}{dt^n}$.

Introducing a differential operator, D such that $D \equiv \dfrac{d}{dt}$, the above equation can be written in a more convenient form:

$$(a_n D^n + \ldots + a_i D + a_o) s_o = (b_m D^m + \ldots + b_1 D + b_o) s_i .$$

It is customary to call the function which relates the time-dependent output signal s_o, to the time-dependent input signal s_i, a "transfer function"[19]. Transfer function relates the amplitude and the phase of the output signal to the amplitude and the phase of the input signal.

Transfer functions may be presented in various forms: operational transfer function, Laplace transfer function, sinusoidal transfer function, constant coefficient, etc.

Presentation of a transfer function in an operational form is very useful. Its modelling, as well as simulation of the related system responses to various input signals, may be easily performed using an analog computer. ˙Such technqiues are being used more and more often.

The concept of transfer function and the derived techniques represent the basis of contemporary analysis and presentation of response of materials and structures, and of performance of loadings and measurement systems, [10,17,19,32-37].

An operational transfer function relating the signal s_o to the signal s_i is obtained by treating the above relation as an algebraic equation,

$$\frac{s_o(D)}{s_i(D)} = \frac{s_o}{s_i}(D) = \frac{b_m D^m + \ldots + b_2 D^2 + b_1 D + b_o}{a_n D^n + \ldots + a_2 D^2 + a_1 D + a_o}$$

where $(s_o/s_i)(D)$ in general depends on time. Laplace and sinusoidal transfer functions are defined in a similar manner.

It is often very convenient to present the operational transfer function in a form of a block diagram:

$$s_i \rightarrow \boxed{F = \frac{b_m D^m + \ldots + b_2 D^2 + b_1 D + b_o}{a_n D^n + \ldots + a_2 D^2 + a_1 D + a_o}} \rightarrow s_o$$

For instance a system represented by the first order form of the basic equation,

$$a_o s_o + a_1 \dot{s}_o = b_o s_i ,$$

may be written in terms of operational transfer function,

$$\frac{s_o}{s_i}(D) = \frac{K}{\tau D + 1}, \text{ where } K = \frac{b_o}{a_o} \text{ and } \tau = \frac{a_1}{a_o},$$

or of block diagram:

$$s_i(t) \rightarrow \boxed{F = \frac{K}{\tau D + 1}} \rightarrow s_o(t) = F[s_i(t)],$$

where the physical sense of parameters K and τ is given by the coefficients

a_o, a_1, b_o.

A function describing the response of a mathematical model to the input signal depends obviously on its form. For instance, when the input signal is given by Heaviside unit step function

$$s_i = A H(t) = s_{is},$$

the response of a system (a body) represented by the first order equation is given by:

$$\frac{s_o}{s_{is}} = K(1 - e^{-t/\tau}).$$

Quantity "τ" determined in this manner is usually called the "time-constant", and is used extensively to characterize the performance of modern measurement systems.

Operational transfer functions are often applied to the analysis of systems which consist of several components, such as the one shown below:

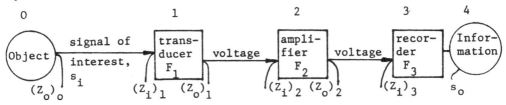

When the output and input impedances are chosen such that all the components of a system respond linearly, for instance when

$$\frac{(Z_o)_{n-1}}{(Z_i)_n} \ll 1,$$

the complete transfer function of the system shown above may be represented by

$$F = F_1 F_2 F_3 \ ,$$

and the system diagram can be reduced to

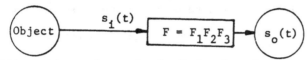

Responses to various standard signals of systems represented by such transfer functions, may easily be analyzed using analog computer techniques.

It is more customary to classify the measurement systems according to their mathematical models, the general form of which is given by the equation (A), page 86 .

A sample of typical classification is given below.

(1) Zero order instrument: $a_n = 0$ and $b_m = 0$ for m,n > 0;
related mathematical model: $a_o s_o = b_o s_i$.

(2) First order instrument: $a_n = 0$ for n > 1; and $b_m = 0$ for m > 0;
related mathematical model: $a_1 \dot{s}_o + a_o s_o = b_o s_i$.

(3) Second order instrument: $a_n = 0$, for n > 2; and $b_m = 0$, for m > 0;
related mathematical model: $a_2 \ddot{s}_o + a \dot{s}_o + a_o s_o = b_o s_i$.

(4) General form of the mathematical model for an nth order instrument:

$$\sum_{k=0}^{k=n} a_k \frac{d^k s_o}{dt^k} = b_o s_i .$$

(5) Dead-time instrument; mathematical model $s_o(t) = Ks_i(t - \tau)$ where

K = constant, and τ = constant delay time.

The same classification can be used to describe responses to various

loads of physical bodies, machines, structures, mechanisms, etc., rep-

resented adequately by particular mathematical models.

6.7   Human Factor. Observer as a Component of a System

It has been accepted in Experimental Mechanics to consider the

experimenter and the observer as elements which are not directly connected

with the system that collects and transfers information. This approach

is changing. There is a strong tendency to consider the human factor,

represented by body senses and the mind, as a part of systems which

produce, collect, transfer and record the information of interest.

Such an approach requires that the response of human senses to input

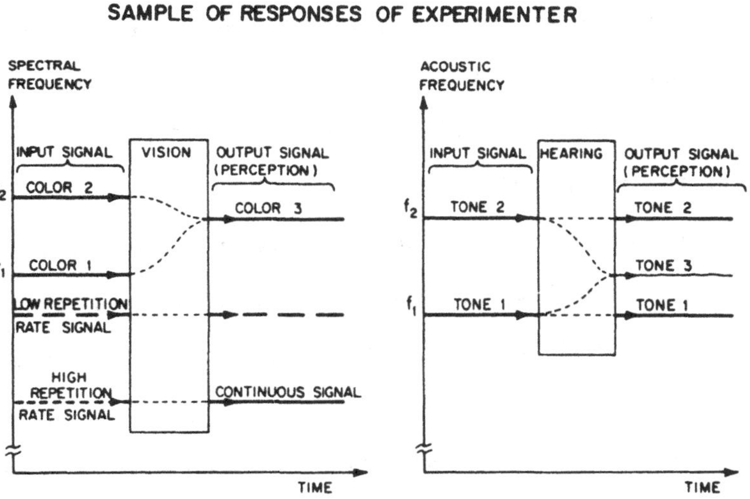

SAMPLE OF RESPONSES OF EXPERIMENTER

Figure 13 - Patterns of response of major senses of experimenter;
            frequency and repetition rate response of vision and hearing.

signals and the transfer of information to consciousness are treated in the same manner as the response of components of any system. This implies that: (1) various input signals can produce the same kind of output signals (perceived information); (2) the same kind of input signal can produce various output signals (reactions, conclusions); (3) the term "illusion" is misleading: it is normal that the transfer function or the main signal can be modified by various modifying signals; (4) knowledge of the transfer functions for signals of interest and knowledge of the major modifying functions of human senses and mind is necessary to optimize (with regard to its reliability) the design of any experiment; and (5) knowledge of environmental parameters which influence processes in the human body is necessary to optimize the design of the experiment with regard to the overall signal/noise ratio.

The following information is of particular interest:

(a)   the spatial and temporal response of hearing as related to the acoustic frequencies, signal amplitude, repetition rate of chopped signals, direction of the path of signal, influence of modifying signals, influence of the signal/noise ratio;

(b)   the spatial and temporal response of vision as related to the spectral frequencies, signal amplitude, repetition rate of chopped signal, geometrical response, influence of modifying signals, influence of the signal/noise ratio;

(c)   the ranges of variation of environmental parameters such as temperature, humidity, acoustic noise level, mechanical vibrations, acceleration (amplitude, frequency and direction), forces acting on body, spectral

distribution of radiant power and intensity of illumination (radiation),
particle radiation (electron, proton, etc.), and especially the knowledge
of: (1) mean values corresponding to optimum efficiency; (2) range of com-
fort; (3) range of tolerance with respect to duration of disturbing signals; (4) range of emergency performance; and (5) range of permanent damage
to human body, and/or its components.

A sample of characteristic responses of human vision and human hearing is given in Figure 13. A typical example of the spacial response of
human vision is given in Figure 14.

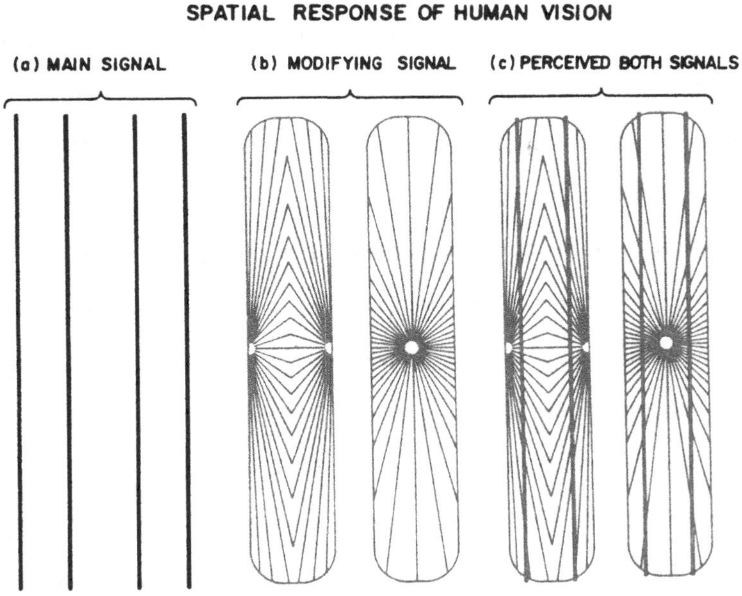

SPATIAL RESPONSE OF HUMAN VISION

Figure 14 - Example of spacial response of human vision:
   (a) main signal is observed;
   (b) modifying signal is observed;
   (c) main and modifying signals are superposed and observed
        simultaneously.

7. SUMMARY OF PART I

7.1 General Summary

In Experimental Mechanics a major criterion for the quality of analytical solution is the degree of correlation between the physical reality and the mathematical model of it on both the theoretical and numerical levels. This degree of correlation between reality and model of it depends on whether or not the chosen approach follows the trend of "physicalisation of mechanics", i.e., whether or not the researcher in his approach:

(a) accepts the idea of the conceptual physical and mathematical models of real phenomena developed on the basis of the correspondence principle; (b) considers the "systems" characterized by "responses" rather than the "isolated processes" characterized by "properties"; (c) accepts the causal-stochastic character of real phenomena, and considers the simple-deterministic approach as a first approximation only; (d) accepts the concept of a discrete structure of matter; (e) accepts the concept of a material continuum, homogeneous and often isotropic, as a useful working hypothesis, with constraints; (f) considers, if desirable, anisotropy as a phenomenon on the molecular level; (g) relates various responses of solids and fluids to their structures; (h) considers the flow of information as a thermodynamic process, related to the flow of energy; and (i) utilizes the concepts of "transfer function" and "impedance" to character- ize directly the flow of energy and indirectly the dynamic responses of simple bodies and complex structures.

On the basis of the presented analysis, one can draw conclusions regarding the methods and techniques of collecting and evaluating experimental results. Obviously, the mean curve presented in Figure 1 represents only the performed experiment and is related to the functional relation between both quantities Q_1 and Q_2 in a manner which depends mainly on the insufficiently known theory of experiment. Figure 1 illustrates the manner of evaluation of the actual relation $Q_1 = Q_1(Q_2)$ by considering the influence of the theory and technique of experiment.

Following the presented relationships and conclusions, one can establish a set of conditions for a reliable evaluation of the degree of correlation between responses of the four major types of iconic models made of matter and energy, Figure 15, responses of mathematical models, and the real phenomenon.

Consequently, one can establish conditions for reliable analysis of results, and one can draw conclusions regarding the structure, the reliability, and the accuracy of predictions of mathematical models which have been developed to simulate the chosen primary patterns of the real phenomena. Patterns of such analysis and of final development of models of reality are presented schematically on Figure 15. This figure represents a continuation and a completion of Figure 5.

The major steps in developing a sufficiently reliable mathematical model may be formulated as: (1) choice of a sufficiently comprehensive conceptual physical model of a phenomenon (2) identification of major parameters influencing the major course of an event within the frame of a chosen physical model, and according to the chosen modelling criteria;

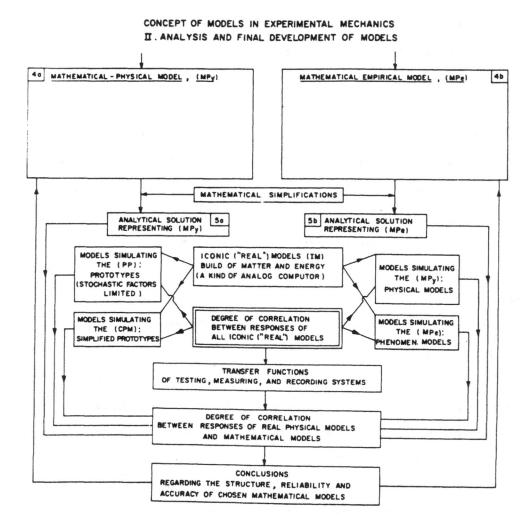

CONCEPT OF MODELS IN EXPERIMENTAL MECHANICS
II. ANALYSIS AND FINAL DEVELOPMENT OF MODELS

Figure 15 - Basic approach to the designing of models in Experimental
            Mechanics: analysis and final stage of model design.

(3) a satisfactory selection of other parameters; and (4) justifiable

analytical simplifications.

        Of course, the term "identification" of major parameters denotes

a choice of relevant parameters of a pertinent, sufficiently general, mathematical model. This problem is illustrated by Figure 10 and the related comments.

7.2   Summary of Modelling Principles in Experimental Mechanics

In real bodies several types of responses can be produced by the same input signal and the same types of response can be produced by various signals.

All responses of real bodies are interrelated. This interrelation makes it possible, for instance, to determine patterns of propagation of elastic waves on the basis of temperature measurements.

Any body under investigation represents a subsystem in an over-all system, part of which is the experimenter. Production, collection and transfer of information implies production of a particular flow of energy which is modulated by the process under study and by all the components of the overall system.

All methods of Experimental Mechanics are developed on the concepts of models of reality, on the basis of principles of conservation of mass and energy, and on the second law of thermodynamics, all of which have been confirmed experimentally. Since any disturbance in real bodies is manifested as a flow of energy, the velocity of propagation of its parti-cular form is the velocity of the related real perturbance. Obviously, any reliable solution in mechanics must comply with these principles.

Practical conclusions are as follows[1-16]:

(a) Any real phenomenon is too complicated to be amenable to rigorous analysis "as it is" and must be represented by suitable models. Models always represent a simplified version of reality and simulate selected processes only. The term model refers to the conceptual models which are described symbolically using words, graphs or sets of analytical expressions, and to the iconic models made of matter and energy; mathematical models represented by graphs or sets of analytical expressions are meaningful only if corresponding sets of conditions and constraints are given. Conceptual models can be speculative (phenomenological), or physical. Speculative models may be related to the real space, or they may be related to a hypothetical space. Physical models are related to the real physical space. As both classes of conceptual models - speculative and physical - can be simulated by iconic models, the physical sense of predictions of the latter depends on the reliability of the former. A given physical process can be represented by various, even disparate, models, related to various frames of reference. For the same reason, a very simplified model can represent various unrelated physical phenomena. The concepts of transfer function which relates the cause to the effect has been developed to give a more convenient analytical tool for analysis of model responses.

(b) Models are invented, not discovered. Models, developed with an objective to represent chosen features of real processes are always invented in a chosen frame of reference related to more general models. The main features of the frame of reference can be geometrical, or can be related to major features of the system under consideration. Of course, the principle of covariance ought to be conserved in all formulations.

(c) Extrapolation of predictions of a model beyond the range of acceptable deviation from the explicitly or implicitly made assumptions usually leads to unnecessary errors. In this respect, it is useful to remember the aphorism "A model is an artifice that gives one the illusion of knowing more about an actual process than one actually does".

(d) Since it is theoretically impossible to obtain information and to perform any investigation without expenditure of energy, energy flow and the related thermodynamical processes represent major phenomena. This energy flow, in turn, influences the process under observation. Any experiment increases entropy of the system. Collection of information increases entropy. Energy flow through the observed object and the measuring system depends on the character of energy transfer between the bodies in contact. It is useful to have a measure for energy transfer between two bodies; as a result the concept of impedance has been developed and recently generalized, and is widely applied.

(e) Since it is impossible to isolate any phenomenon being investigated, it is necessary to identify all major factors which can significantly influence the process under investigation. This leads to the concepts of "system", "system response", "noise" and of "modifying signals".

(f) Concepts such as "properties of materials" represent only a simplified manner of describing the parameters of functional relationships for materials, a manner which is rooted in the concept of "true" mathematical models. These relationships are influenced by the shape of a specimen, by the method of investigation, by all the parameters of the energy flow, and by all the components of the testing, measuring and recording systems. Usage of such terms in the sense of intrinsic quantities may lead to misunderstandings, and therefore the term "property" is understood in Experi-

mental Mechanics as an idiom rather than in the sense of an intrinsic
quantity; the pertinent rigorous terms are "response" or "behaviour".

Consequently, in Experimental Mechanics every theory ought to be
related to real physical world and must comply with the foundations of
physics and with major theories of physics. The intellectual and utili-
tarian interest in speculative theories, based on assumed physical reality
must be recognized[23].   Preference, however, is given to theories which
comply with the accumulated knowledge regarding physical reality, when any
conflict between both kinds of theories exists.

As a result, in Experimental Mechanics the solutions are analysed
with respect to the actual process, not with respect to other assumed
mathematical models. "Rigorous solution" means a solution which is rigorous both mathematically and physically, that is, which represents all major
physical events occurring in the phenomenon under investigation. Hence,
a distinction is made between "reliability" as related to the degree of
correlation between a phenomenon and its mathematical model (including
the complete measuring system), and "accuracy" as related to repeatability
of measurements under identical conditions. The latter encompasses the
classical terms accuracy and precision and the related definitions.

Reliability of a mathematical model depends on its compatibility
with the corresponding physical theories and the actual physical processes
(a theoretical problem). Numerical accuracy of predictions of a mathematical model is to be analysed with respect to the real process, and not
with respect to another mathematical model. According to this condition
a rigorous mathematical solution of a mathematical model which was con-

structed in an empirical manner represents an empirical solution of a
problem, regardless of the rigorousness of mathematical transformations.

Obviously, no measuring instrument indicates or records stress,
strain, acceleration, birefringence, etc. Only the power or energy arriv-
ing at the measuring instrument can be displayed or recorded. Consequently,
experimental results are always presented within a particular frame of
reference given by a chosen mathematical model of the phenomenon itself,
and of the involved system: an "objective presentation" does not exist.
Thus to analyse the reliability and accuracy of experimental results it is
necessary to know the responses of the whole system which influences the
patterns of flow of energy through it. System responses can be presented
by transfer functions of subsystems, and by impedances which determine
the exchange of energy between subsystems.

7.3   Main Areas of Interest in Experimental Mechanics

The following main areas of contemporary Experimental Mechanics may
be outlined.

(a) Determination of the major course of event and major parameters which
characterize the response of physical bodies to mechanical, thermal and
other loads. The primary objectives are to decide which quantities should
be considered as major parameters, and to obtain functional relations
between them and input signals. Subsequently, it is necessary to investigate to what extent the shape of specimen and the response of the whole
system influences the indicated response of material to loads. Finally,
the most interesting problem is to establish relations between the various
responses.

(b)   Analysis of mathematical models developed in analytical and numerical
mechanics with regard to the conceptual and numerical correlations between
these models and the real processes. This area of research is of particular importance since the solutions of concrete problems are often based
on assumptions which do not reflect satisfactorily real processes. The
main issues are: (1) degree of correlation between the chosen mathematical
model and the major course of event of the process under consideration
(conceptual correlation); and (2) degree of correlation between responses
predicted by the mathematical models and the actual functional relations
between major parameters of the actual process (numerical correlation).

(c)   Development of new mathematical models of material and structural
responses, or development of major elements of such mathematical models.
The need for research in this area arises when too great a discrepancy is
observed between the prediction of mathematical models and the real response of element. This is presently the case regarding viscoelasticity,
thermal phenomena related to transient elastic states, vibrations, contact problems, etc.

(d)   Development of new theories, methods and experimental techniques for
studies of responses of engineering materials and structures, and for
measuring the quantities of interest. There is a strong interaction in this
area of activity with research in the fields of experimental physics,
materials science, information theory, control theory, data processing
techniques, etc. Activity in this area and directions of development are
strongly influenced by the economical factors and by the level of education and technical training of persons who are the potential users of

methods, techniques and instrumentation.

(e) Development of theoretical foundations of standardized testing

techniques. Regarding the current and predicted interest of industry, and

technology in general, this is one of the major fields of activity in

experimental research.

(f) Integration and condensation of knowledge. There is a growing

discrepancy between the level of sophistication regarding the theories

and techniques of experimental methods being developed and applied, and

the level of teaching of related subjects in the institutions of higher

learning. The principle formulated in the sentence:

> "do not start any experiment until you
> understand the theory behind it"

cannot be followed if the institutions of higher learning do not provide

adequate intellectual and professional background, or if the industrial

organizations do not support the additional necessary training. In spite

of the seriousness of this growing problem there is still a tendency to

neglect it.

(g) Experimental design optimization of new structures within given

conditions and constraints. These activities, beside rendering requested

technological services, assure that the necessary link is maintained

between research and practice.

7.4   Conclusions

The main general objective of research in Experimental Mechanics is

to develop mathematical models of phenomena, within a framework of a

mathematical model of the whole system, which are characterized by a

satisfactory degree of correlation with real phenomena and by satisfactory

accuracy of predictions regarding the chosen parameters of response. The
rapid development of numerical methods utilizing computer techniques
emphasizes the importance of this objective. The present trend is clear.
Numerical methods of stress analysis are replacing gradually in several
areas the methods and techniques of experimental stress analysis. However, the reliability of mathematical models assumed in the numerical
methods of mechanics, and the accuracy of predictions of these methods
can be determined experimentally only. Cooperation of Experimental
Mechanics with analytical and numerical mechanics with regard to the
development of mathematical models seems to be one of the major issues in
the future development in applied mechanics,[38-44]

The ultimate application of new theories, methods and techniques of
Experimental Mechanics is in Industrial Research and Development, to attain
design optimization with regard to the strength, service life, stiffness,
frequency spectrum, physiological parameters of the response of human
body, etc., under imposed economical, social and ecological conditions
and constraints.

PART II: EXAMPLES OF PRACTICAL CONSEQUENCES OF
PHYSICAL APPROACH

8. COUPLED RHEOLOGICAL RESPONSES OF MATERIALS USED IN MODEL MECHANICS

Experimental evidence presented in[45-67, 71] shows that:

(1) Major responses of iconic model materials are coupled; the mechanism of coupling is often complicated.

(2) Creep of all typical polymers, used for iconic and photoelastic models, with the exception of glass at room temperature, is not negligible, when quantitative measurements are conducted.

(3) Linear range of mechanical and optical rheological responses of major photoelastic materials is limited and depends on time.

(4) Stress- or strain-induced birefringence is a nonmonotonic function of wavelength of electromagnetic radiation, both in linear and non-linear ranges; consequently, the wavelength of radiation is a major parameter of a photoelastic experiment.

(5) Stress-induced birefringence (photoelastic effect) is seldom caused by one mechanism only; in general it is due to two, or three, or more effects, even at room temperature.

(6) Orientational effects in materials, on molecular levels, are primarily responsible for the optical viscoelastic and inelastic effects.

(7) It is not possible to understand the rheological response of all photoelastic materials, with the possible exception of crystals, without pertinent information on the related creep-recovery function.

(8) Photoelastic effect resulting from superposition of static and dynamic

oscillating loads cannot be described and explained by a linear
model.

(9) Photo-thermal effects due to oscillating kinematic loading (deformations prescribed) are strongly nonlinear.

(10) Only an isochronous presentation of the tensile curve allows for
unequivocal assessment of the mode of viscoelastic deformation,
both linear or nonlinear.

The results presented in [50, 58, 59] are compatible with some results
presented in earlier papers, for example [47] which describes the influence of the orientational birefringence produced by styrene molecules;
however, they cannot be predicted by the available phenomenological
models of viscoelasticity.

8.1   Proposed Measures of Coupled Responses

On the basis of information presented in pertinent bibliography,
and illustrated by Figures 16-28, the following quantities, which will be
discussed subsequently, appear to be of major importance either as quantities complementary to the set of typical rheological functions or
quantities necessary for a satisfactory designing of these functions,
Ref. [24-30, 49-51, 54-60, 71].

(1) Variation of the thermal expansion coefficient with the temperature
and the related variation of mechanical damping: quantities describing the character of influence of temperature on the response
of materials, Figures 16 and 17.

(2) Spectral transmittance: quantity describing the related interaction
between radiant energy and matter, Figure 18.

(3) Creep recovery and relaxation recovery functions: represent an extension of the common creep compliance functions and the relaxation functions into the creep recovery and the relaxation recovery regions, Figures 19, 25 and 26.

(4) Isochronous stress-strain, and stress-birefringence relations: functions indicating linearity or nonlinearity of time-dependent responses, Figures 20 and 21.

(5) Linear limit stresses $\sigma_{\ell\ell}$, the corresponding linear limit strains $\varepsilon_{\ell\ell}$ and linear limit birefringence $\Delta n_{\ell\ell}$: quantities describing the ranges of the linear rheological response, Figures 22, 23 and 28.

(6) Normalized spectral birefringence: quantity describing the dependence of the magnitude of relative retardation on spectral frequency of radiation, Figure 24.

(7) Spectral dispersion of optical axes: quantity describing the dependence of the angle between chosen mechanical principal directions and optical principal directions, on the frequency of radiation, Figure 27.

(8) Normalized spectral dispersion of birefringence: quantity describing the rate of change of the spectral birefringence with the radiation frequency.

(9) Normalized dielectric coefficient: quantity describing the dependence of the dielectric coefficient on the field frequency.

The actual rheo-optic behaviour of several polymer materials, described by the above formulated measures, can be understood on the basis of a more comprehensive physical model of the interaction between radiation and matter, some details of which are given in Table 2.

Table 2.   Sample of interaction between radiation and matter at the
atomic and molecular levels.

spectral region / quantity	\multicolumn{7}{c}{Spectral regions of electromagnetic radiation}						
	microwaves	far infrared	infrared	near infrared	visible	ultra violet	x-rays
wave length — millimeters	10·0	1·0 0·1	0·05 0·01	2×10⁻³ 1×10⁻³			
wave length — microns	10,000	1000 100	50 10	2 1	0·74 – 0·38	0·1	
wave length — nanometers				2000 1000	740 – 380	100	
frequency — Giga-Hertz (10⁹ Hz)	30	300 3000	30×10³	3×10⁵		3×10⁶	
wave number — cm⁻¹	1·0	10·0 100·0	10³	10⁴		10⁶	
photon energy — e.v.	125×10⁻⁶	125×10⁻⁵ 125×10⁻⁴				125	
photon energy — erg	2×10⁻¹⁶	2×10⁻¹⁵ 2×10⁻¹⁴					
nature of radiation		Rotational spectra. Long-wavelength vibrational spectra	Vibrational- rotational spectra	Long-wave length electronic spectra. Vibrational-rotational spectra	Outer-shell electron transitions	Inner shell electron transitions	
present photoelastic research	⊢—⊣			⊢———⊣	⊢═══⊣ ⊢—⊣		

8.2   Details of Proposed Measures

8.2.1   Rheological Functions: Recovery Functions, Compliance Functions.

The mechanical and optical creep recovery functions carry information
that is not only supplementary to the information provided by the creep
function, but also supplies essential information on the mechanism of
deformation. Obviously only the creep and creep recovery functions which
are produced by a step load function are amenable to simple analysis
without additional assumptions.

Experimental results shown in Figures 19–27, can be explained
satisfactorily only by assuming that rheo-birefringence is produced by a
set of partial effects, Figure 28. While the particular shape of the

creep function can be explained as a result of two partial effects, one distortional and one orientational, creep recovery function indicates that there exists at least two orientational birefringence effects, produced at different stress levels and related to different time intervals.

Typical mathematical models of the rheological response of several high polymers are not commensurate with the real response of various major groups of polymers, such as polyester and epoxy resins, in both linear and nonlinear ranges. Creep compliance functions designed to represent the creep part of creep response, in both the linear and nonlinear ranges cannot describe satisfactorily the response of some materials, Figures 19 and 25.

8.2.2   Linear Limit Stresses. The isochronous relations between stress, strain and birefringence are linear in limited range or ranges. The response of materials can also be linear in one or more ranges, as presented schematically in Figures 20 and 21.

It is useful to introduce measures for the linear ranges of the rheological response of materials in the form of linear limit stresses, $\sigma_{\ell\ell}$ and linear limit strains, $\varepsilon_{\ell\ell}$, Figure 22.

Two particular measures have been introduced: the nominal linear limit stress $(\sigma_{\ell\ell})_{nom}$ or $\sigma_{\ell\ell}$, and the percental linear limit stress $(\sigma_{\ell\ell})_{\%}$. Both measures can be related to the primary, $\sigma'_{\ell\ell}$, or to the secondary, $\sigma''_{\ell\ell}$, linear ranges of response.

The concept of the nominal limit stress is of interest with regard to the mathematical models of material response. The percental linear limit stress, which often describes the width of the transition range between the two linear ranges, is also of engineering interest.

The linear limit stresses and strains are time-dependent; they may
be different for the strain and the birefringence, Figure 23. They also
depend on temperature and decrease to zero in the ranges of transition
temperatures. Further, the linear limit stresses for birefringence may
depend on the wavelength of radiation.

For generality, such relations should be presented as follows:
for strain: $(\sigma_{\ell\ell})_\varepsilon = \sigma_{\ell\ell}(t,T)$, and for birefringence: $(\sigma_{\ell\ell})_{\Delta n} = \sigma_{\ell\ell}(t,T,\lambda)$.

The dependence of the linear limit stress on time, temperature and
wavelength of radiation supplies information on the mechanism of deformation.

In accordance with the interpretation of experimental results, given
in this discussion, the multi-linear response of materials, presented
schematically on Figures 22 and 28, can be approximated by the relation

$$\varepsilon(t), \; \Delta n(t) = D_o\sigma + \sum_k D_k(t)W_k(T) <\sigma - \sigma_{\ell\ell}(t)_k>^1 \, ,$$

where $W_k(T)$ denotes the statistical weight which is temperature dependent,
D_o denotes the instantaneous value of the creep compliance function, D_k
denotes the partial creep compliance functions representing partial effects,
and $(\sigma_{\ell\ell})_k$ denotes the linear limit stress of the kth partial effect.

8.2.3 Normalized Spectral Birefringence. The dependence of components
of the index tensor on frequency of electromagnetic radiation is one of
the major features of interaction between radiation and matter. Consequently, the absolute stress-induced birefringence and all the derived
quantities must depend on spectral frequency. Experimental results show
that relative birefringence, and the derived quantities depend on spectral
frequency as well.

It is convenient to represent the spectral frequency dependence of birefringence by introducing the normalized spectral birefringence, r, defined as follows:

$$r = \frac{R\,(\lambda)}{R\,(\lambda_o)} = \frac{\Delta n(\lambda)}{\Delta n(\lambda_o)} = r(\lambda) = kr(\nu),$$

where R denotes the relative retardation in terms of the distance between two retarded wave fronts, Δn denotes the relative retardation in terms of the difference between the principal indices of refraction related to both wave fronts, λ and ν denote the wavelength and spectral frequency of radiation, respectively, and k is a transformation factor.

In the linear range of response for t = const.:

$$r(\lambda) = \frac{C_\sigma(\lambda)}{C_\sigma(\lambda_o)} = \frac{C_\varepsilon(\lambda)}{C_\varepsilon(\lambda_o)} .$$

Since the dependence of the indices of refraction on spectral frequency is a fundamental feature of the interaction between radiation and matter which depends directly on the structure of matter and the spectral frequency, spectral birefringence should be considered as a basic measure of the mechanism of birefringence and of mechanism of rheological processes.

Experimental results confirm the above interpretation. Figure 24 presents the influence of the manner of producing stress birefringence on the spectral birefringence. The birefringence produced in glass by the residual stresses induced by a suitable rheological process is different from the birefringence produced by force-induced stresses. The difference between the two spectral birefringence functions is qualitative in the near infrared band of radiation. In this case spectral birefringence is a suitable measure of the rheological past. Similarly the relations presented in Figure 25 show that spectral birefringence is related to different mechanisms of deformation below and above transition temperature.

However, G.L. Cloud has shown that for some polymers spectral birefringence at constant temperature, normalized with respect to time, does not depend on spectral frequency in a wide spectral band,[27].

8.2.4   Normalized Spectral Dispersion of Birefringence.   In accordance with common practice, spectral dispersion of birefringence is defined as:

$$D = \frac{dr}{d\lambda} = \frac{1}{\Delta n(\lambda_o)} \frac{d(\Delta n(\lambda))}{d\lambda} = D(\lambda).$$

In the linear range of response for t = const.:

$$D(\lambda) = \frac{1}{C_\sigma(\lambda_o)} \frac{dC_\sigma(\lambda)}{d\lambda} = \frac{1}{C_\varepsilon(\lambda_o)} \frac{dC_\varepsilon(\lambda)}{d\lambda}$$

It is convenient to normalize this quantity:

$$d(\lambda) = \frac{D(\lambda)}{D(\lambda_o)}.$$

The normalized spectral dispersion of birefringence is a convenient method of presenting the rate of change of spectral birefringence with respect to wavelength, and subsequently with respect to time and to temperature.   According to experimental evidence this quantity is closely related to the character of rheological deformation, both linear and nonlinear.

It has been shown that the spectral dispersion of birefringence in a birefringent liquid is a certain measure of linear response[71].   — The spectral range where the value of spectral dispersion of birefringence is minimum coincides with the spectral range where the linear range of birefringence response to shear strain rate is maximum.

8.2.5 Other Measures. Other quantities which characterize linear and
nonlinear mechanical, optical and electrical responses of materials of
interest in Experimental Mechanics, as discussed in [55-60] are:
normalized dielectric coefficient; spectral dispersion of optical axes;
spectral transmittance; thermal expansion coefficient; transition
temperature.

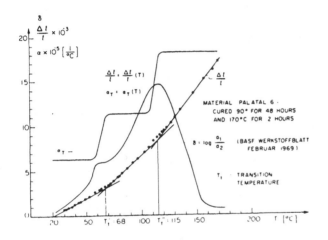

Fig. 16 A sample of
coupled responses: the
transition temperatures
are indicated by
changes of three para-
meters of response.

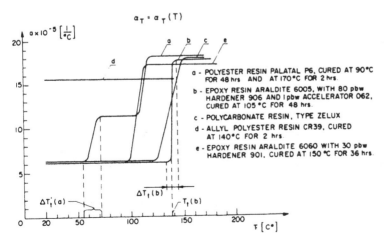

Fig. 17 Temper-
ature dependence
of linear thermal
expansion coef-
ficients of some
model materials

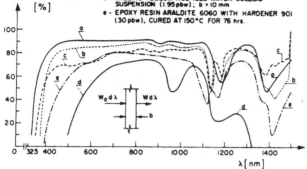

Fig. 18 - Spectral transmittance of several photoelastic materials.

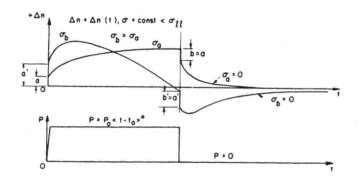

Fig. 19 - Typical mechanical and optical creep and creep recovery responses to a limited unit step load, spectral frequency dependent.

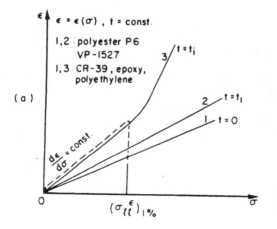

Figure 20 - Typical isochronous stress-strain responses.

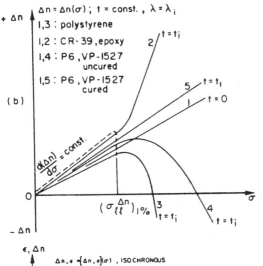

Fig. 21 - Typical isochronous stress-birefringence responses

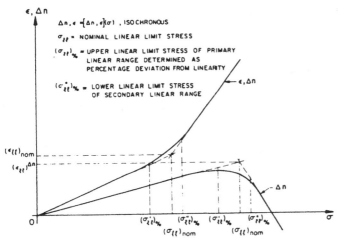

Fig. 22 - Concept and definition of the nominal and per- cental limit stresses and strains for strain and bire- fringence.

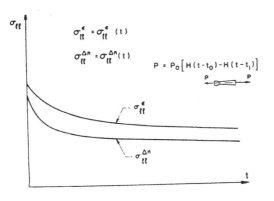

Fig. 23 - Typical time dep- endence of linear limit stresses for strain and bire- fringence below transition temperature.

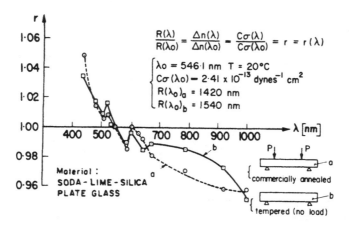

Figure 24 - Sample of normalized spectral birefringence: dependence
on history of deformation.

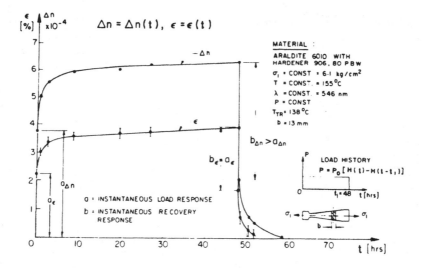

Figure 25 - Mechanical and optical creep and creep recovery responses
above transition temperature.

$$D = D(t) ; \quad C_\sigma = C_\sigma(t) ; \quad C_\epsilon = C_\epsilon(t)$$

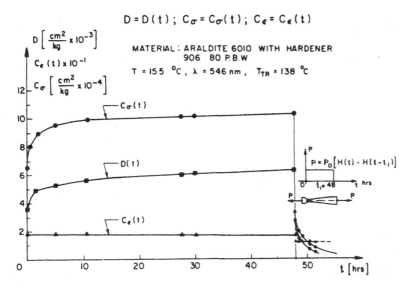

MATERIAL : ARALDITE 6010 WITH HARDENER
906   80 P.B.W

$T = 155\ °C$, $\lambda = 546\ nm$, $T_{TR} = 138\ °C$

Fig. 26 – Time
dependence of
mechanical,
stress–optical
and strain-
optical rheo-
logical
functions, above
transition
temperature.

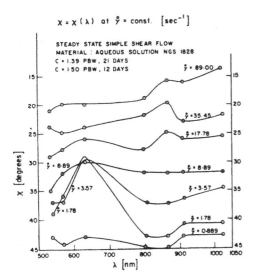

$\chi = \chi(\lambda)$ at $\dot{\gamma} = const.$ $[sec^{-1}]$

STEADY STATE SIMPLE SHEAR FLOW
MATERIAL : AQUEOUS SOLUTION NGS 1828
C = 1.39 PBW, 21 DAYS
C = 1.50 PBW, 12 DAYS

Figure 27 – Example of the
spectral dependence of
optic axes.

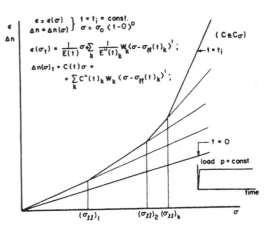

$$\left.\begin{array}{l} \epsilon = \epsilon(\sigma) \\ \Delta n = \Delta n(\sigma) \end{array}\right\} \begin{array}{l} t = t_i = const. \\ \sigma = \sigma_0 \langle t - 0 \rangle^0 \end{array}$$

$$\epsilon(\sigma)_t = \frac{1}{E(t)}\sigma + \sum_k \frac{1}{E''(t)_k} W_k \langle \sigma - \sigma_{\ell\ell}(t)_k \rangle^1 ;$$

$$\Delta n(\sigma)_t = C(t)\sigma + \sum_k C''(t)_k W_k \langle \sigma - \sigma_{\ell\ell}(t)_k \rangle^1 ;$$

Figure 28 – Schematic presentation
of behaviour of a viscoelastic
material exhibiting two linear
ranges connected by a wide transition
region.

9. STRESS STATE IN A CIRCULAR DISK-CONTRADICTIONS BETWEEN THE ANALYTICAL
    AND EXPERIMENTAL RESULTS

Circular disk loaded by compressive forces applied at the ends of
one diameter belongs to a small group of elastic bodies for which elegant
analytical solutions have been available for a long time,[78-81], and for
which the boundary conditions are easily reproducible experimentally. As
a result such disks are extensively used to assess the reliability and
accuracy of photoelastic techniques,[81]. For a long time such disks
have been used as the so-called "creep compensators" or "creep calibrators";
they allow to easily account for the optical and mechanical creep, temperature influence and influence of the wavelength of the detecting radiation
used, to reliably evaluate the momentary values of the stress optic coefficient at the time of recording,[82-84].

The classical analytical solution, based on the known stress function
predicts that along the loaded diameter the stress component normal to it
$\sigma_x(0,y)$, is constant and positive. Because of this unusual feature of
the stress field and the simple shape and simple loading conditions, some
standardized tensile tests for brittle materials are using such circular
disks or circular cylinders as standard specimens. In particular such a
test is used in various countries to determine the tensile strength of
concrete and its validity is widely discussed[85-87].

Since the concentrated loads produce high strain and stress gradients
and noticeable elastic deformation in the regions of load application, the
question is how reliable is the analytical solution within those regions,
and what is the influence of eventual deviation on the whole strain and
stress fields in the disk.

A modified theory developed by G. Hondros [85] predicts that the value of the above stress component may vary along the loaded diameter and even alter the sign, when the loading force is distributed over a small area. According to this theory $\sigma_x(0,y)$ alters its sign at the distance of 0.85R from the centre of the disk, when the loading force is distributed over the angle of 0.042 rad. with respect to the disk centre.

The experimental results reported in [86] and [87] correlate very well with each other and correlate well with the analytically determined values of stress components and their functions in the regions far away from the points of the load application, Figure 29. The experimentally determined stress field in the region within the distance of 0.20R - 0.25R from the points of the load application deviates qualitatively and quantitatively from the classical analytical predictions: not only the values of stress components differ but their sign differs, too. Around the regions of load application the stress components in the plane of the disk are compressive.

One can conclude that:

- The classical theory predicts satisfactorily the stress components in the disk in the regions sufficiently remote from the points of load application. However, the stress field in the region up to the distance of ca. 0.3 disk radius from the points of load application cannot be determined reliably by the classical theory.

- It appears desirable to reassess the validity of all tensile tests for brittle materials based on the classical disk theory.

- It seems necessary to introduce two characteristic points of the plane stress field in disk loaded diametrically and specify their locations

on the loaded diameter:

(1) two reversal points: $\sigma_x(0,y) = 0$,

(2) two singular points: $\sigma_1 - \sigma_2 = 2\tau_{max}(0,y) = 0$.

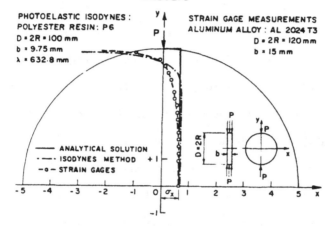

NORMAL STRESS DISTRIBUTION $\sigma_x(0,y) = \sigma_1(0,y)$
COMPARISON BETWEEN THE ANALYTICAL AND EXPERIMENTAL
RESULTS

Figure 29 - Ranges of reliable predictions of classical analytical
         solution for stress state in a circular disk loaded
         diametrically, indicated by values of stress components
         obtained analytically and experimentally.

10. NON-RECTILINEAR LIGHT PROPAGATION IN A STRESSED BODY

10.1 Introduction.

Many experimental methods use light as a detector and/or carrier

of information. The related theories of experiments and experimental

techniques are founded on the assumption that the light propagation

within the transparent bodies is always rectilinear.

This assumption is not always justified. It is known that the

light path is not rectilinear when light travels through a body density of which, and the related index of refraction, vary   from point to point.

The density alteration may or may not be related to anisotropy of the

body - a body can be non-homogeneous and at the same time isotropic or

anisotropic - in both cases the light path will be bent. This effect

limits the resolution of the classical photoelastic observations, as

shown by M.F. Bokshtein,[88].

An example of the light path curvature caused in a rectangular

prismatic specimen by a residual (frozen) birefringence is shown in Figure

30, Ref.[83].The recordings made in diffused light represent typical

recordings of a transparent specimen, made using the natural and polarized

light, respectively. .However, the recordings made using collimated light

seem to indicate that the faces of the specimen were curved; a lense

effect is observed. The explanation of this effect is given in Figure 31,

Ref. [83]; light path is bent towards regions of higher density.

The same effect of the curvature of light path in a stressed glass

sample has been observed by Acloque and Guillemet[89]   and utilized to

determine residual stress state in glass. P. Manogg applied this effect

in fracture mechanics,[90].

The above described effect is, in principle, well known under the

name "Schlieren" and has been used for a long time to detect local

inhomogenities caused mainly by the local density alteration, which, in

turn, produce local alteration of values of refractive index.

Basic theories of the described effect are given in all modern

textbooks on optics,[91].   However, in engineering research it was

believed that the density alteration caused by the symmetrical part of

LIGHT PROPAGATION THROUGH SOLID BODY
RECORDINGS OF OPTICAL EFFECTS

SPECIMEN: BEAM 88 x 12 mm. MADE OF CATALIN 800
WITH RESIDUAL STRESSES

☀ NATURAL LIGHT, WHITE   ⓓ DIFFUSED LIGHT

Ⓘ CIRCULARLY POLARIZED   ⓒ COLLIMATED LIGHT
LIGHT   λ = 546 nm   SMALL APERTURE

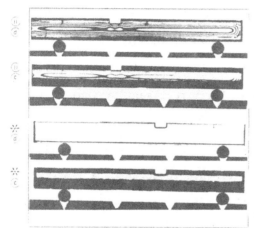

Fig. 30 (above): Optical effects
caused by curvature of light
path

LIGHT PROPAGATION THROUGH A SOLID BODY:
INFLUENCE OF INHOMOGENEITY GRADIENT, AND
ANISOTROPY GRADIENT CAUSED BY STRAIN/STRESS
GRADIENTS

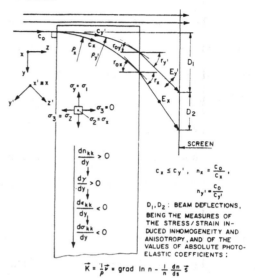

$$\frac{dn_{kk}}{dy} > 0$$

$$\frac{dy}{dy} > 0$$

$$\frac{d\epsilon_{kk}}{dy} < 0$$

$$\frac{d\sigma_{kk}}{dy} < 0$$

$$c_x \leq c_{y'}, \quad n_x = \frac{c_o}{c_x},$$

$$n_{y'} = \frac{c_o}{c_{y'}}$$

D_1, D_2: BEAM DEFLECTIONS,
BEING THE MEASURES OF
THE STRESS/STRAIN IN-
DUCED INHOMOGENEITY AND
ANISOTROPY, AND OF THE
VALUES OF ABSOLUTE PHOTO-
ELASTIC COEFFICIENTS:

$$\vec{K} = \frac{1}{\rho}\vec{\nu} = grad \ ln \ n - \frac{1}{n}\frac{dn}{ds}\vec{s}$$

LIGHT PROPAGATION THROUGH SOLID BODY. DEFLECTIONS
IN FUNCTION OF MEAN STRESS GRADIENTS

$$\frac{1}{2}(\Delta D_x + \Delta D_y) = f_s\left(\frac{dS}{dy}\right), \quad S = \frac{\sigma_x + \sigma_y}{2}$$

MATERIAL: POLYESTER RESIN P6

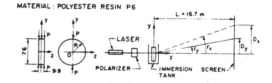

Fig. 31 (above right); Influence
of gradients of plane stress
field on light path

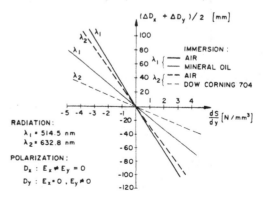

RADIATION:
$\lambda_1 = 514.5$ nm
$\lambda_2 = 632.8$ nm

POLARIZATION:
$D_x : E_x \neq E_y = 0$
$D_y : E_x = 0, E_y \neq 0$

IMMERSION:
λ_1 { —— AIR
       —— MINERAL OIL
λ_2 { –– AIR
       –– DOW CORNING 704

Fig. 32 (right): Gradient
photoelasticity: dependence
of light deflection on gradient
of the sum of principal
stresses.

the stress or strain tensor and the anisotropy caused by the deviatoric part are too insignificant to impair the accuracy of optical measurements or to produce any measurable effect. Available evidence[83,88-90], was disregarded.

It has been shown recently[92-94] that the effect of the unavoidable curvature of light path in photoelastic specimens is twofold:

- it limits the resolution of transmission photoelastic measurements;
- it supplies data on gradients of the sum and difference of principal stresses, which represent foundation of gradient photoelasticity[95].

Gradient photoelasticity supplies complementary data on plane stress fields and yields values of absolute photoelastic coefficients, Figure 32.

The reliability and accuracy of results depend on the reliability and accuracy of the transfer function of the measurement system[29,33-35,56].

11. ISODYNE PHOTOELASTICITY

The method of photoelastic isodynes utilizes the modulation of intensity of radiation scattered at points of a plane within a body, when the scattered radiation satisfies the following conditions at each scattering point: direction of primary beam is parallel to chosen axis; angle of observation is equal to $\pi/2$; azimuthal angle with respect to the unperturbed primary beam is constant and close to 0 or $\pi/2$.

The geometric loci of points of constant light intensity are called isodynes[96] because they represent lines of constant value of total normal force acting on the cross-section corresponding to the related section of the primary beam,

$$I_s^x (x,y) = \text{const} = kP_y = S_s m_s,$$

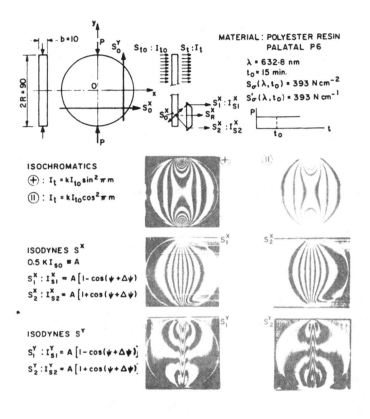

ISOCHROMATICS

$\oplus$: $I_t = kI_{t_0}\sin^2\pi m$

$\odot$: $I_t = kI_{t_0}\cos^2\pi m$

ISODYNES S^x

$0.5\,KI_{so} = A$

$S_1^x : I_{s_1}^x = A\left[1 - \cos(\psi + \Delta\psi)\right]$

$S_2^x : I_{s_2}^x = A\left[1 + \cos(\psi + \Delta\psi)\right]$

ISODYNES S^Y

$S_1^Y : I_{s_1}^Y = A\left[1 - \cos(\psi + \Delta\psi)\right]$

$S_2^Y : I_{s_2}^Y = A\left[1 + \cos(\psi + \Delta\psi)\right]$

Fig. 33 – Isodyne
photoelasticity:
basic relations and
samples of
recordings.

STRENGTH OPTIMIZATION USING INTEGRATED ISODYNE TECHNIQUE

$\lambda = 632.8\,nm$ $S_\sigma(\lambda, t) = 383\,Ncm^{-2}$ $S_\sigma'(\lambda, t) = 383\,Ncm^{-1}$

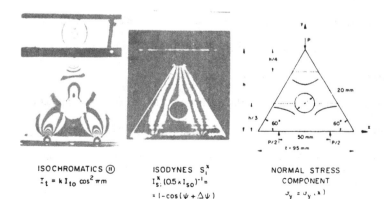

ISOCHROMATICS $\odot$

$I_t = kI_{t_0}\cos^2\pi m$

ISODYNES S_1^x

$I_s^x (0.5\,\kappa I_{so})^{-1} =$

$= 1 - \cos(\psi + \Delta\psi)$

NORMAL STRESS
COMPONENT

$\sigma_y = \sigma_y(x)$

Fig. 34 – Example
of determination
of stress
components using
isochromatics and
isodynes only.

when this beam propagates through a plane body, parallel to its surface

in the x-direction; the slope of the isodyne order diagram with respect

to the direction of the primary beam is proportional to the related

normal stress component, [96-99]. Consequently,

$$\sigma_y = S_\sigma \, (dm_s/dx).$$

Figures 33 and 34 illustrate this method. It should be noted that

it is necessary to know the transfer function of the isodyne polariscope

to evaluate correctly the stress components [34].

12. LIMITS OF CONVENTIONAL ULTRASONIC TECHNIQUES

Ultrasonic techniques utilize elastic waves excited in an object by

a suitable transducer. The conventional ultrasonic techniques applied

for crack sizing are based on measurements of the transit time and ampli-

tude of the returned signal.

The distance of a crack from the surface can be determined by

measuring the transit time of the signal from the transducer to the crack

and back; the location of the crack can be determined when the position

of the ultrasonic transducer is known. The size of the crack is deter-

mined by measuring the amplitude of the returned signal; this technique

is based on the assumption that the returned signal is a reflected signal.

These conventional techniques are all based on the concepts and

relations of geometric optics, according to which, for instance, the

relation between the crack size and the echo amplitude should be un-

equivocal. Figures 35 and 36 [100], shows that the geometric optics cannot

provide a necessary theoretical background of sufficiently reliable and accurate ultrasonic techniques for crack sizings. It appears that the echo amplitude is significantly influenced by some interference and diffraction effects, which can be successfully analyzed by means of wave theory only[101].

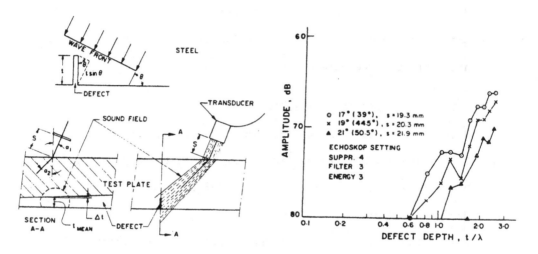

Figure 35 - Ultrasonic crack sizing; scheme of measurement

Figure 36 - Dependence of echo amplitude on crack depth.

REFERENCES

1. Brillouin, L., *Scientific Uncertainty, and Information*, Academic Press, New York 1964.

2. Popper, K.R., *The Logic of Scientific Discovery*, Harper and Row, London and New York, 1959, (1968).

3. Krajewski, W., *Correspondence Principle and Growth of Science*, D. Reidel Publishing Company, Dordrecht-Holland, Boston, U.S.A., 1977.

4. Kuhn, T.S., *The Structure of Scientific Revolution*, University of Chicago Press, Chicago, Illinois, 1962, 1970.

5. Wade, N. and Kuhn, T.S., Revolutionary Theorist of Science, *Science*, 197, 143-145, 1977.

6. Sedov, L.I., *Introduction to the Mechanics of a Continuous Medium*, (translated from the Russian edition, Moscow, 1962), Addison-Wesley Publishing Co., 1965.

7. Sedov, L.I., On Prospective Trends and Problems in Mechanics of Continuous Media, (in Russian), *Prikladnaya Matematika i Mekhanika*, 40, 963-980, 1976. English translation, Pergamon Press Ltd., 1977.

8. Tiller, A.W., Materials Science and Applied Science, *Science*, 16, 469-475, 1969.

9. Kac, M., Some Mathematical Models in Science, *Science*, 166, 695-699, 1969.

10. Pindera, J.T., Contemporary Trends in Experimental Mechanics: Foundations, Methods, Applications, in *Experimental Mechanics in Research and Development*, Pindera, J.T., Leipholz, H.H.E., Rimrott, F.P.J. and Grierson, D.E., Eds., SM Study No. 9, University of Waterloo Press, Ontario, Canada, 1973.

11. Olszak, W., Mechanics Today: Lights and Shadows, Lecture given at Canadian Congress of Applied Mechanics, Fredericton, May 26-80, 1975.

12. Sedov, L.I., Theoretical Constructions of Selection of Actual Events from the Virtual Ones, in manuscript; courtesy of the author, 1979.

13. Pindera, J.T., How General can be Teaching of Experimental Stress Analysis, *Proc. of the Fourth Int. Conf. on Experimental Stress Analysis*, Cambridge, April 6-10, 1970, Institution of Mechanical Engineering, London, 542-545, 1971.

14. Pindera, J.T., Problems of Reliability of Common Models of Basic Responses of Materials and Systems, *Proc. of the VIII Symposium on Experimental Research in Mechanics of Solids*, September 4-6, 1978, Warsaw, Poland.

15. Hussey, M., *Modelling I and II*, The Open University Press, Walton Hall, Bletchley Buckinghamshire, 1972.

16. Naughton, J., *Scientific Method and System Modelling*, The Open
    University Press, Walton Hall, Milton Keynes, 1975.

17. Pindera, J.T., Leipholz, H.H.E., Rimrott, R.P.J. and Grierson, D.E.,
    Eds., *Experimental Mechanics in Research and Development*, Proc. of
    the Int. Sym. held at the University of Waterloo, June 12-16, 1972,
    University of Waterloo Press, Solid Mechanics Division, Waterloo,
    Ontario, Canada, 1973.

18. Axelrad, D.R., *Random Theory of Deformation of Structured Media*
    and Axelrad, D.R. and Provan, J.W., *Thermodynamics of Deformation
    in Structured Media*, International Centre for Mechanical Sciences,
    Udine, Italy, Springer-Verlag, New York, 1972.

19. Doeblin, E.O., *Measurement Systems: Application and Design*,
    McGraw-Hill, 1975.

20. Soodak, H. and Iberall, A., Homeokinetics: A Physical Science for
    Complex Systems, *Science*, 201, 579-582, 1978.

21. Alexrad, D.R., *Micromechanics of Solids*, Elsevier Scientific
    Publishing Co., Amsterdam and New York, 1978.

22. Reiner, M., Rheology, in *Encyclopedia of Physics*, S. Flügge, Ed.,
    VI, 434-510, Springer-Verlag, Berlin, 1958.

23. Leipholz, H.H.E., Analytical Foundations of Experimental Mechanics.
    Trends in Analytical Mechanics, in *Experimental Mechanics in
    Research and Development*, Pindera, J.T. et al, Eds., Solid Mechanics
    Division, University of Waterloo Press, Ontario, Canada, 1973.

24. Van Geen, R., Dispersion chromatique de l'effet photoelastique,
    Proc. 2nd Int. Conf. on Experimental Stress Analysis, Paris, France,
    April 10-14, 1962.

25. Pindera, J.T. and Cloud, G.L., On Dispersion of Birefringence of
    Photoelastic Materials, *Experimental Mechanics*, 6, 470-480, 1966.

26. Cloud, G.L. and Pindera, J.T., Techniques in Infrared Photoelasticity,
    Experimental Mechanics, 8, 193-201, 1968.

27. Cloud, G.L., Mechanical Optical Properties of Polycarbonate Resin
    and Some Relations with Material Structure, *Experimental Mechanics*,
    9, 489-499, 1969.

28. Pindera, J.T. and Straka, P., On Physical Measures of Rheological
    Responses of Some Materials in Wide Ranges of Temperature and Spectral
    Frequency, *Rheologica Acta*, 13, 338-351, 1974.

29. Pindera, J.T. and Sinha, N.K., On the Studies of Residual Stresses
    in Glass Plates, *Experimental Mechanics*, 11, 113-120, 1971.

30. Pindera, J.T., Alpay, S.A. and Krishnamurthy, A.R., New Developments
    in Model Studies of Liquid Flow by Means of Flow Birefringence,
    Trans. of the CSME, 3, 95-102, 1975.

31. Kestin, J.A., *Course in Thermodynamics*, Vol. I and II, Blaisdell
    Publishing Co., Toronto, 1966, 1968.

32. Sciammarella, C.A., Basic Optical Law in the Interpretation of Moiré
     Patterns Applied to the Analysis of Strains - Part 1, *Experimental
     Mechanics*, 5, 154-160, 1965.

33. Pindera, J.T., On the Transfer Properties of Photoelastic Systems,
     in *Proc. of the Seventh All-Union Conf. on Photoelasticity*, Tallinn,
     November 23-26, 48-63, 1971.

34. Pindera, J.T. and Straka, P., Response of Integrated Polariscope,
     Journal of Strain Analysis, 8, 65-76, 1973.

35. Pindera, J.T., Response of Photoelastic Systems, *Trans. of the CSME*,
     2, 21-30, 1973-74.

36. Mayer, N. and Rohrbach, C., *Handbook for Fluidic Measurements*,
     (in German), VDI-Verlag, Düsseldorf, 1977.

37. Stein, P.K., *Measurement Engineering, Vol. I: Basic Principles*,
     6th Edition, Stein Engineering Services, Inc., Phoenix, Arizona,
     1970.

38. Lewicki, B. and Pindera, J.T., Photoelastic Models of Reinforced
     Structures, (in Polish), *Archiwum Inzynierii Ladowej*, 2,   381-418,
     1956.

39. Pindera, J.T. and Sze, Y., Studies of Physical and Mathematical
     Models of Some Flanged Connections, in *Proc. of the Fourth Int.
     Conf. on Experimental Stress Analysis*, April 6-10, 1970, Cambridge,
     England, Institution of Mechanical Engineering, Westminster, London,
     395-408, 1971.

40. Pindera, J.T. and Sze, Y., Response to Loads of Flat-Faced Flanged
     Connections and Reliability of Some Design Methods, *Trans. of the
     CSME*, 1, 37-44, 1972.

41. Pindera, J.T. and Sze, Y., Characteristic Parameters of Response
     of Plates in Contact, in: *Proc. of the 2nd Int. Conf. on Structural
     Mechanics in Reactor Technology*, Berlin, September 10-14, 1973,
     Paper M5/8, 1-12.

42. Pindera, J.T. and Sze, Y., Influence of the Bolt System on the
     Response of the Face-to-Face Flanged Connection, in: *Proc. of
     the 2nd Int. Conf. on Structural Mechanics in Reactor Technology*,
     Berlin, September 10-14, 1973, Paper G2/6, 1-13.

43. Pindera, J.T., Experimental Study of Some Problems Related to Res-
     ponses of Thick Plates, (in German), in: *Experimentelle Spannungs-
     und Dehnungsanalyse*, Laermann, K.-H., Ed., Verner-Verlag, Düsseldorf,
     25-49, 1977.

44. Laermann, K.-H., *Experimental Investigations of Plates. Theoretical
     Foundations*, (in German), Verner-Verlag, Düsseldorf, 1977.

45. Stuart, H.A., *Physics of High Polymers*, (in German), 4, Springer-
     Verlag, Berlin, 1956.

46. Jira, T., Mechanical and Photoelastic Behaviour of Celluloid at
     Biaxial Load, (in German), *Konstruktion*, 9, 438-449, 1957.

47. Hiltscher, R., Theorie and Application of Photoelasticity in Elastic-
     Plastic Range, (in German), *VDI Zeitschrift*, 97, 49-58, 1955.

48. Loreck, R., Investigation of Suitability of Polyester Resin "Leguval"
     and Some Other Polymers as Materials for Photoelastic Models, (in
     German), *Kunststoffe*, 52 , 139-143, 1962.

49. Pindera, J.T., Some Research Work in Photoelasticity Carried Out in
     the Polish Academy of Sciences, (in Russian), in: *Polarization-
     noopticheski metod issledovania napriazheni Trudy Konferentsi*,
     February 13-21, 1958, Izd. Leningradskogò Universiteta, 32-44, 1960.

50. Pindera, J.T., Rheological Properties of Some Polyester Resins,
     Part I, II and III, (in Polish), *Rozprawy Inzynierskie*, 3, 361-411,
     481-540, 1959.

51. Pindera, J.T., Some Rheological Problems at Photoelastic Investiga-
     tions, in *Proc. Int. Spannungsoptisches Sym.*, Berlin, April 10-15,
     1961, Akademic-Verlag, Berlin, 155-172, 1962.

52. Read, B.E., Dynamic Birefringence of Amorphous Polymers, *Journal
     of Polymer Science*, Part C, 87-100, 1964.

53. Ward, I.M. and Pinnock, P.R., The Mechanical Properties of Solid
     Polymers, *British Journal of Applied Physics*, 17, 3-32, 1966.

54. Pindera, J.T., Remarks on Properties of Photoviscoelastic Model
     Materials, *Experimental Mechanics*, 6 , 375-380, 1966.

55. Pindera, J.T. and Kiesling, E.W., On the Linear Range of Behaviour
     of Photoelastic and Model Materials, *Proc. Third Int. Conf. on
     Experimental Stress Analysis*, Berlin, 1966, VDI-Berichte, No. 102,
     VDI-Verlag, Düsseldorf, 89-94, 1966.

56. Pindera, J.T., On Physical Basis of Modern Photoelasticity Tech-
     niques, *Berträge zur Spannungs-und Dehnungsanalyse*, Vol. V, Academie-
     Verlag, Berlin 103-130, 1968.

57. Kiesling, E.W. and Pindera, J.T., Linear Limit Stresses of Some
     Photoelastic and Mechanical Models Materials, *Experimental Mechanics*,
     9, 337-347, 1969.

58. Pindera, J.T. and Straka, P., On Physical Measures of Rheological
     Responses of Some Materials in Wide Ranges of Temperature and
     Spectral Frequency, *Rheologica Acta*, 13, 338-351, 1974.

59. Pindera, J.T., Straka, P. and Krishnamurthy, A.R., Rheological
     Responses of Materials Used in Model Mechanics, in: *Proc. of the
     Fifth Int. Conf. on Experimental Stress Analysis*, held in Udine,
     Italy, May 27-31, 1974, CISM, Udine, 2.85-2.98, 1974.

60. Pindera, J.T., Straka, P. and Tschinke, M.R., Actural Thermoelastic Response of Some Engineering Materials and its Applicability in Investigations of Dynamic Response of Structures", *VDI-BERICHTE*, 313, 579-584, 1978.

61. Andrews, R.D. and Hammack, T.J., Temperature Dependence of Orientation Birefringence of Polymers in the Glassy and Rubbery States, *Journal of Polymer Science*, Part C, Polymer Symposia, Stein, R.S., Ed., No. 5, Interscience Publishers, 101-112, 1964.

62. Maxwell, J.C., Double Refraction of Viscous Fluids in Motion, *Roy. Soc. Proc.*, 22, 46, 1873-74.

63. Kundt, A., On the Birefringence of Light in Moving Viscous Liquids, (in German), *Wiedmann's Annalen*, XIII, 110, 1881.

64. Natanson, M.L., O pewnej właściwości podwojnego załamania światła w cieczach odkształcanych mogącej posłużyć do wyznaczania ich czasu zluźniania, (Sur une particularité de la double rèfraction accidentelle dans les liquides pouvant servir à la détermination de leur temps de relaxation), *Bull. Acad. Sci.*, Cracovie, 1-22, 1904.

65. Zaremba, S., Note sur la double refraction accidentelle de la lumière dans les liquides, *J. de Phys.*, 3, 606-611, 1904 and 4, 514-516, 1905.

66. Zakrzewski, K., O położeniu osi optycznych w cieczach odkształcalnych (Sur la position des axes optiques dans les liguides deformés), *Bull. Acad. Sci.*, Cracovie, 50-56, 1904.

67. Wiener, O., Laminal Birefringence, (Lamellare Doppelbrechung), *Physikalische Zeitschrift*, 5, 332, 1904.

68. Roman, C.V. and Krishan, K.S., A Theory of the Birefringence Induced by Flow in Liquids, *Phil. Mag.*, 5, 769-783, 1928.

69. Peterlin, A. and Stuart, H.A., Uber den Einfluss der Rotationsbehinderung und der Anisotropie des inneren Feldes aug die Polarisation von Flüssigkeiten, *Z. Phys.*, 113, 663-696, 1939.

70. Philipoff, W., Flow Birefringence and Stress, *J. Appl. Physics*, 27, 984-989, 1956.

71. Pindera, J.T. and Krishnamurthy, A.R., Characteristic Relations of Flow Birefringence, Part 1: Relations in Transmitted Radiation, *Experimental Mechanics*, 18, 1-10, 1978, Part 2: Relations in Scattered Radiation, *Experimental Mechanics*, 18, 41-48, 1978.

72. Jerrard, H.G., Theories of Streaming Double Refraction, *Chem. Rev.*, 59, 345, 1959.

73. Wayland, M., Streaming Birefringence of Rigid Macromolecules in General Two-Dimensional Laminar Flow, *J. Chem. Phys.*, 33, 769, 1960.

74. Mindlin, R.D., A Mathematical Theory of Photo-Viscoelasticity, *J. Appl. Phys.*, 20, 206-216, 1949.

46. Jira, T., Mechanical and Photoelastic Behaviour of Celluloid at Biaxial Load, (in German), *Konstruktion*, 9, 438-449, 1957.

47. Hiltscher, R., Theorie and Application of Photoelasticity in Elastic-Plastic Range, (in German), *VDI Zeitschrift*, 97, 49-58, 1955.

48. Loreck, R., Investigation of Suitability of Polyester Resin "Leguval" and Some Other Polymers as Materials for Photoelastic Models, (in German), *Kunststoffe*, 52, 139-143, 1962.

49. Pindera, J.T., Some Research Work in Photoelasticity Carried Out in the Polish Academy of Sciences, (in Russian), in: *Polarization-noopticheski metod issledovania napriazheni Trudy Konferentsi*, February 13-21, 1958, Izd. Leningradskogo Universiteta, 32-44, 1960.

50. Pindera, J.T., Rheological Properties of Some Polyester Resins, Part I, II and III, (in Polish), *Rozprawy Inzynierskie*, 3, 361-411, 481-540, 1959.

51. Pindera, J.T., Some Rheological Problems at Photoelastic Investigations, in *Proc. Int. Spannungsoptisches Sym.*, Berlin, April 10-15, 1961, Akademic-Verlag, Berlin, 155-172, 1962.

52. Read, B.E., Dynamic Birefringence of Amorphous Polymers, *Journal of Polymer Science*, Part C, 87-100, 1964.

53. Ward, I.M. and Pinnock, P.R., The Mechanical Properties of Solid Polymers, *British Journal of Applied Physics*, 17, 3-32, 1966.

54. Pindera, J.T., Remarks on Properties of Photoviscoelastic Model Materials, *Experimental Mechanics*, 6, 375-380, 1966.

55. Pindera, J.T. and Kiesling, E.W., On the Linear Range of Behaviour of Photoelastic and Model Materials, *Proc. Third Int. Conf. on Experimental Stress Analysis*, Berlin, 1966, VDI-Berichte, No. 102, VDI-Verlag, Düsseldorf, 89-94, 1966.

56. Pindera, J.T., On Physical Basis of Modern Photoelasticity Techniques, *Berträge zur Spannungs-und Dehnungsanalyse*, Vol. V, Academie-Verlag, Berlin 103-130, 1968.

57. Kiesling, E.W. and Pindera, J.T., Linear Limit Stresses of Some Photoelastic and Mechanical Models Materials, *Experimental Mechanics*, 9, 337-347, 1969.

58. Pindera, J.T. and Straka, P., On Physical Measures of Rheological Responses of Some Materials in Wide Ranges of Temperature and Spectral Frequency, *Rheologica Acta*, 13, 338-351, 1974.

59. Pindera, J.T., Straka, P. and Krishnamurthy, A.R., Rheological Responses of Materials Used in Model Mechanics, in: *Proc. of the Fifth Int. Conf. on Experimental Stress Analysis*, held in Udine, Italy, May 27-31, 1974, CISM, Udine, 2.85-2.98, 1974.

75. Green, A.E., Rivlin, R.S. and Spencer, A.J.M., The Mechanics of
     Non-Linear Materials with Memory, Part I, II and III, *Archiv
     Rational Mechanics Anal.*, 1, 1-21, 1957; 3, 82-90, 1959 and 4,
     387-404, 1960.

76. Coleman, B.D., Dill, E.H. and Toupin, R.A., A Phenomenological
     Theory of Streaming Birefringence, *Arch. Rational Mech. Anal.*, 39,
     358-399, 1971.

77. Theocaris, P., Phenomenological Analysis of Mechanical and Optical
     Behaviour of Rheo-Optically Simple Materials, in: *The Photoelastic
     Effect and its Applications*, Kestens, J., Ed., Springer-Verlag, 1975.

78. Mushelishvili, N.I., *Some Basic Problems of the Mathematical Theory
     of Elasticity*, P. Noordhoff, Groningen-Holland, 324-328, 1953.

79. Sokolnikoff, I.S., *Mathematical Theory of Elasticity*, McGraw-Hill
     Book Company, Toronto, 283-287, 1956.

80. Timoshenko, S.P. and Goodier, J.N., *Theory of Elasticity*, McGraw-
     Hill Book Company, Toronto, 122-127, 1970.

81. Frocht, M.M., *Photoelasticity, Vol. II*, John Wiley, New York,
     121-129, 1948.

82. Pindera, J.T., *Outline of Photoelasticity*, (in Polish), P.W.T.,
     Warszawa, 1953.

83. Pindera, J.T., Technique of Photoelastic Studies of Plane Stress
     States, (in Polish), *Rozprawy Inzynierskie*, Polish Academy of
     Sciences, 3, 109-176, 1955.

84. Pindera, J.T., *Contemporary Methods of Photoelasticity*, (in Polish),
     Panstwowe Wydawnictwa Techniczne, Warszawa, 1960.

85. Hondros, G., The Evaluation of Poisson's Ratio and the Modulus of
     Materials of a Low Tensile Resistance by the Brazillian Test,
     Australian J. of Appl. Science, 10, 243-268, 1959.

86. Pindera, J.T., Mazurkiewicz, S.B. and Khattab, M.A., Stress Field
     in Circular Disk Loaded Along Diameter: Discrepancies Between
     Analytical and Experimental Results, presented at the *SESA Spring
     Meeting*, Wichita, Kansas, May, 1978, Paper No. CR-10.

87. Chong, Ken P., Finite Element and Other Analyses of Split Disks,
     in manuscript, 1978.

88. Bokshtein, M.F., On Resolving Power of Photoelastic System for
     Stress Analysis, (in Russian), *J. Tekhn. Fiziki*, XIX, 1103-1106,
     1949.

89. Acloque, P. and Guillemet, G., Method for the Photoelastic Measure-
     ment of Stresses "In Equilibrium in the Thickness" of a Plate,
     (Particular Cases of Toughened Glass and Bent Glass), *Selected
     papers on Stress Analysis presented at the Institute of Physics,
     Stress Analysis Group Conference*, Delft, 1959.

90. Manogg, P., The Light Deflection in an Elastically Deformed Plate
     and the Shadow Patterns of Circular Notches and Cracks, (in German),
     Glastechnische Berichte, 39, 323-329, 1966.

91. Born, M. and Wolf, E., *Principles of Optics*, 5th Edition, Pergamon
     Press, Oxford, New York, 1975.

92. Hecker, F.W. and Pindera, J.T., Influence of Stress Gradient on
     Direction of Light Propagation in Photoelastic Specimens,
     VDI-BERICHTE, 313, 745-754, 1978.

93. Hecker, F.W., Kepich, T. Y. and Pindera, J.T., Neglected Factor in
     Photoelasticity: Non-linear Light Propagation in Stressed Bodies
     and Its Significance, *Proc. of The 8th All-Union Conf. on Photoelas-
     ticity*, Tallinn, September 25-28, 1979, Akademia Nauk Estonskoy
     SSR, Institut Kibernetiki, Tallin, 1, 117-123, 1979.

94. Hecker, F.W., Kepich, T.Y.   and Pindera, J.T., Non-Rectilinear
     Optical Effects in Photoelasticigy Caused by Stress Gradients, in:
     Proc. IUTAM Sym. on Optical Methods in Mechanics of Solids, Poitiers,
     September 10-14, 1979.

95. Pindera, J.T., Hecker, F.W. and Krasnowski, B.R., A New Experimental
     Method: Gradient Photoelasticity, to be published.

96. Pindera, J.T. and Mazurkiewicz, S.B., Photoelastic Isodynes: A New
     Type of Stress Modulated Light Intensity Distribution, *Mech. Res.
     Comm.*, 4, 247-252, 1977.

97. Mazurkiewicz, S.B. and Pindera, J.T., Integrated-Plane Photoelastic
     Method-Application of Photoelastic Isodynes, *Experimental Mechanics*,
     19, 225-234, 1979.

98. Pindera, J.T., Elements of More Rigorous Theory and Technique of
     Isodyne Method and Their Applications to Other Optical Methods, in:
     Proc. IUTAM Sym. on Optical Methods in Mechanics of Solids, Poitiers,
     September 10-14, 1979.

99. Pindera, J.T. and Mazurkiewicz, S.B., Optimization of Photoelastic
     Stress Analysis using Isodyne Method, in: *Proc. of The 8th All-
     Union Conf. on Photoelasticity*, Tallinn, September 25-28, 1979,
     Akademia Nauk Estonskoy SSR, Institut Kibernetiki, Tallinn, I, 145-
     150, 1979.

100. Hartung, H.F., Burns, D.J. and Pindera, J.T., Ultrasonic Monitoring
     of Growth of Part-Through Thickness Defects at 290°C, *Trans. ASME,
     Journal of Engineering for Power*, 101, 471-476, 1979.

101. Kino, Gordon S., Nondestructive Evaluation, *Science*, 206, 173-180,
     1979.

102. Roe, P.H., Soulis, G.N., Handa, V.K., *The Pincipline of Design*,
     Printed in Canada at the University of Waterloo, 1969.

103. Dixon, J.R., *Design Engineering*, McGraw-Hill, New York, 1966.

FLOW VISUALIZATION

by

Christian TRUCHASSON
Professeur sans chaire (I.N.P. - ENSEEIHT)
2, rue Charles Camichel
31071 TOULOUSE CEDEX - France

In first, we present what kind of information we can hope from flow visualization and some generalities about it.

Roughly we shall classify the methods of flow visualization into four groups. In the first class (chapter 2) visible foreign material is added to the gazeous or liquid flowing fluid : one observes the motion of the foreign material, called "tracer", in the fluid.

In the second class (chapter 3) the fluid elements are discriminated by an energetic injection instead of material injection.

The third class (chapter 4) exploits the index variation without injection, and can be ragarded as undisturbing methods.

In the fourth class (chapter 5) we present analogical methods : flow visualization is made with a different fluid instead of the primitive fluid required by the original study.

In last chapter (6) we present the technical helps in photography of flow.

From immemorial days, people have been fascined by natural shows including flow visualization : waves in the sea and on the rocks, clouds running in the sky, trunks carried by rivers, dust and sleeves flying in the air, lava over flowing from a volcano.

Technical progress added bubbles on loW pressure regions of marine propeller, mist in wake of wing-planes, chemical and industrial dye in the air and in rivers, petrol carpets on the ocean.

The first reaction is often fear, afterwards, the curiosity became interest and all these visual sensations gave a kind of knowledge due to custom. In the same time, these visual people sensations are the source of scientific evidence that any observer could control.

In the beginning, natural or controlled flow-visualization played an important role in the understanding of fluid mechanics problems.

Nowadays, a scientist studying a nature scale flow or a model scale flow is always pleased by the show given by a visualization, even when other quantitative measurements are needed, and even if more precise knowledge must be improved about stability, permanency, turbulence, pressure and velocity distribution.

Seeing is the beginning of understanding. Flow visualization is always helpful and sometimes sufficient for solving engineer problems.

Because most fluids are transparent media, in many cases their motion remains invisible to the human eye in direct observation. A flow visualization technique is to make visible informations about the flow that would stay invisible without this technique. If possible, quantitative data should be obtained from flow picture.

Specially in laboratory, the authenticity of the information is very important : the scale effect has to be studied, and the chosen technique must not disturb the structure of the flow. When using tracers, their material in mass and geometry, the injective process, and the investigation technique have to be criticized and the error observation or measurement have to be evaluated.

The flow visualization technique is often a disturbing method with affects the original flow according to the amount of released energy or material. Nevertheless some of them will be regarded as undisturbing methods.

1. GENERALITIES

In flow visualization, the observer must understand the meaning of the show. The main information of which seems to be the velocity distribution. The velocity vector is tangent to streamlines and filament lines and pathlines that must be distinguished by an observer, in an unsteady flow.

1.1. Velocity information

A particle path is the material line followed by a particular fluid particle in the flow. It is easy to get particle paths with solid tracers (2.2) ; it is possible with coloured liquid (2.1) or bubbles injection (2.4) in a liquid.

A filament line or streakline is the material line made of all the particles which have passed through a geometrically fixed point of the flow. A point permanent injection shows a filament line. Dye lines in

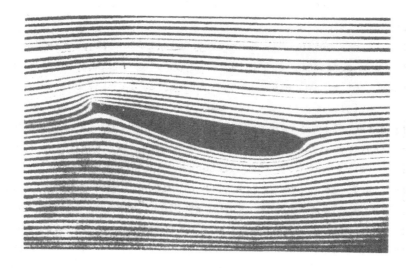

LINES IN A PERMANENT FLOW

liquid fluid flows (2.1) give the filament line that could also be obtained by smoke injection in air (2.2) or by spark tracers (3.3) or cheminu-
minescence (3.5)

A streamline at the instant t_o is a geometrical line each point of
which has the instantaneous velocity vector for tangent. Because of the
need of a chronophotographic technique the streamlines can be get in some
cases of material tracer (2) of energetical injection (3) by mean of
combination of the injective technique and the camera shutter speed.

The shape of all this curves changes with the velocity of the
observer : each line is different when it is recorded with a moving
camera or with a camera at rest in the laboratory system. For instance,
an unsteady wave flow may appear steady to an observer who moves at the
wave velocity with the flow : one must be cautious and thoughtful in
interpreting flow photographs of streamlines in rotative machine.

The differentiation between streamlines, filament lines, and
particles paths disappears in steady flows.

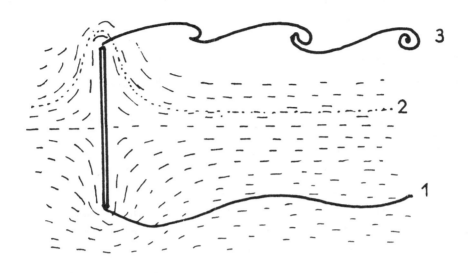

PATH-LINE (1), STREAMLINE AT T_0 (2) AND FILAMENT LINE (3)
IN UNSTEADY FLOW

1.2. Other informations

Flow visualization gives direct information about the pressure in
a part of the flow by means of vapor bubbles in cavitative liquid flow
(2.6).

Through the index variations, all the undisturbing optical methods
(4) give information about the density variations and the shock waves
position. The electron beam technique (3.4) can also be used. By the way
of optical and energetical injection part of the flow with high density
gradient (or high temperature gradient) can be achieved.

Diffusion problems, melange problems and the boundary layer of a
profile can be studied by dye line (2.1), smoke line (2.2), electrolytic
dye production (2.5). The shape, the thichness, the transition points
and the shape of the wake can be studied by many ways, but specially
with the dye injection in fluid (2.1), undisturbing optical methods (4)
and streaming double refraction (5.1). This method is also very convenient

to study the stability, the steadiness and the laminar or turbulent
character of the flow ; material injection (2) and energetical injections
(3) can be used too.

Sometimes, like vegetals in a gust of wind, the flow visualization
shows surface effect by mean of tufts or soft paints, specially for aeronautical purposes (2.6).

2. MATERIAL TRACER TECHNIQUES

When a foreign material is added to a flowing fluid, one may
assume that the flow of the fluid is not disturbed and that, in direction
and in magnitude of velocity, the motion of these visible particles is
the same as that of the fluid itself. But it is only an assumption, and
the difference between the fluid and the particle motion cannot be
avoided, but must be minimized by giving to the particles a density
almost coinciding with that of the fluid, and a size as small as possible. That is the reason why mosquitos smashed on the front glass of a
running car.

Nevertheless, in liquid fluid flows, coloured tracers can give
spectacular and faithful results.

2.1. Coloured liquid in liquid flow

A dye solution can be prepared, for instance with alcohol, to get
the same specific weight as the working fluid, generally water. Werle
(1960) by mixing dye with milk obtained beautiful results, in combination with the hydrodynamic analogy. Food coloured milked dye presents
the advantage of retarding the diffusion into the neighbourhood and of
giving a high contrast of the lines. It is easy to photograph coloured
filament lines. One can get a good idea of the velocity along such a
streakline with a camera and a pulsed injection of dye. Flow and pressure
in the injector must be carefully tested and controlled. The injective
problem from a small injection in the flow or from small wall holes can
alter the main flow of the boundary layer mass and momentum transfer
and even change laminar into turbulent layer.

For seeing in red the turbulent part of boundary layer and in green
the laminar part, Werle and Gallon (1969) used a dye solution the colour
of which changes with concentration.

A fluorescent dye in water has been used in the beginning of fluid
mechanics research, and in nature for source identification. It is a
good mean for improving the visibility of the filaments with an adequate
radiation of the light and an adaptive sensibility of the camera film.

A review of the techniques in water is given by Clayton and Massey
(1967) from fluorescent rhodamine, to methylene blue dye, black nigrosin,
gentian violet, crystal violet, printer's process white, and food colou-
ring and milk.

A chemical time reaction, described by denbigh (1962) was used by
Danckwerts and Wilson (1963) for visualization of reversing flow regime :
by choosing the reaction time for the mixture to turn blue, the visualization shows the "lazy fluid" of the flow. The mixture contained
$Na_2S_2O_8$ and starch.

Another chemical help in flow visualization may occur by mean of
electrolytic dye production (2.5).

Of course, all this popular methods of visualization are restricted
to a tunnel without close-circuit, because of the coloured liquid pollution. It is the same with solid tracers.

2.2. Solid tracers

Smoke can visualize the flow pattern of a gas stream like a dye
solution in a liquid. Brown (1953) refined this kind of old technique
applied to scientific experiments by Mach (1896).

The smoke should be dense and white for visibility, non toxic in
open tunnel, should not distrub the experimental flow conditions, have no
important solid deposit and give good photographs.

Among the best compromises, there are a paraffin mist generator,
titanium-tetrachloride, filtered smoldering cellulose and smoke obtained
by cooking various materials (Malby and Keating 1962)..Beautiful teaching
movies have been got by this way (Educational Services Incorporated ;

Encyclopedia Britannica Educational Corporation).

An optical arrangement and a phototube enabled Becker (1967) to measure turbulent concentrations fluctuations by testing the concentration of particles in a control volume.

Many years ago, dust in water in combination with chronophotographic equipment (C. Camichel 1919) gave quantitative measurement of the velocity in laboratory experiments and aluminium flakes floating on a channel gave famous results (Prandtl).

Many tracer particles can be used, according to the fluid and to the purpose (cf Merzkich 1974).

VISUALIZATION WITH BLACK INK SOLUTION IN OIL
(C. FONADE, I.M.F. Toulouse)

The choice of the solid particle tracer is governed by a high light reflectivity, by the density of the particle, its dimensions smaller than the fine structure of the flow under study. Theoretical approach of the choice problem has been attempted.

2.3. The theoretical problem of particle trajectories and the choice of particles

Because the particles used in flow visualization are not of the same size and density than the main fluid molecules, there can be differences between particle and fluid motion.

If the flow presents a velocity gradient, there is a lift force parallel to the gradient (cf inviscid solution of potential flow). Theoretical calculation gave the results that smaller the size, more faithful is the trajectory, and that lift force is very smaller than drag force when the velocity gradient is not too strong. Measuring the flow close to a solid wall requires very small particles ($\phi < 1\mu$m) for minimizing the lift force.

In compressible fluid flow, the lift and the drag coefficients are fonction of Raynolds and Mach numbers (Crowe 1967). In shock waves, because of heat and momentum transfer, the particles need a relaxation time to get the true velocity of the fluid flow (Stein and Pfeifer 1972).

The charge of the flow velocity in turbulent fluctuations is a problem.

The helium bubble flow visualization is available. The technique employs small, neutrally buoyant bubbles to describe paths and velocities in an air flow.

The particle injection device should not interfere with the flow or should be located for enough upstream from the test field.

Generally, the observer requires that in visualization field, the particle sinks downward less than their diameter.

In fact, the selection of particles is determined by the quality of the recording system, the dimensions of the flow field and its charateristics.

Among the bubble tracers in liquid, the hydrogen bubble technique has known a large expansion in free surface flow because it allows a close-loop experimentation.

2.4. The hydrogen bubble technique

Sometimes gas bubbles are distributed at random in a fluid and can be used for flow visualization.

In cavitation flow, vapor bubbles form and grow as a consequence of pressure reduction : the bubbles collapse as they are swept downstream into the high-pressure region. Cavitation conditions can be observed in the cores of the tip vortices from the blades of a marine propeller, in the flow past a disc in the zones of high turbulent shear at the edges of a separated wake or in the low pressure core of the laminar boundary layer separation region. Stroboscopic lighting or high speed motion pictures reveal the flow in the low pressure region.

Gas bubbles can be generated by means of the electrolysis of the fluid if an electrolytic solution. Flow visualization is possible with tap water by the hydrogen bubble technique ; it is very attractive because the bubbles can be produced at a controlled rate and a any desired position in the flow. One point gives a strekline if bubbles are generated steadily ; a fine wire cathode with pulsing voltage at a constant frequency gives time lines and velocity profile ; if the cathode wire has short sections coated with insulation and is normal to the mean flow we get a time-streak marker technique very efficient for teaching and exhibition.

THE HYDROGEN BUBBLE TECHNIQUE USED FOR MAPPING A STEADY FLOW
(from E.B.C. Catalogue)

The time of the observation, limited by the dissolution of the bubbles in water, is about 3 sec, but in turbulent flows owing rapid diffusion of the bubbles. In water, velocity must be smaller than 30 cm/sec and wire Reynolds number must be laminar.

Laminar wake theory gives that the bubble velocity reaches free stream velocity about 70 wire diameters downstream : the test regime in the flow should be at least 70 diameters behind the wire.

Because a gas bubble can change its shape during the motion, the drag coefficient of such a tracer particle is a function of the velocity difference between the fluid ans the tracer, and of the deforming forces; a quantitative analysis of the bubble motions is complicated and the buoyancy forces in the flow superpose a vertical rise velocity of which the associated errors can be considerable if the flow field contains steep vertical velocity gradients. If the bubbles can be generated very small in size, the Reynolds number and the velocity of rising motion is low and can be determined by Stokes'law.

The bubble motion is photographed to get measurement. The total error in the measured velocity seems to be of the ordre of 5 % because of wake velocity, averaging process and measuring imperfections of a careful experimentation.

Such experiments of films carried out are very appropriate for fluid mechanics teaching.

An obvious advantage of this method is its non pollution quality so that it can be applied in recirculating water channels.

2.5. Electrolytic dye production

Buoyancy effects do not exist in thymol blue dye. The time streak technique can be applied. This method is appropriate for convective flows, flows in stratified fluids, rotating flows and pulsating flows because of the same density of the tracer particles.

The rate of diffusion increasing with the mean velocity, limits the investigation to the range of 0 to 5 cm/s : at higher speeds and in turbulent wake the edges of the tracer become diffuse. The Reynolds

numbers of cathode wire are often very low and limit the allowable velocity (Re   40). A corrective analysis gives a multiplying factor
depending of the time interval between electric impulse and photographic
recording.

Like in hydrogen bubble technique the electrode wire wakes give
experimental errors.

In tellurium method in water, the cathode is a thin tellurium wire
under about 300 volt direct current giving about 1 A/cm of cathode wire
length (Wortmann). Tellurium ions with a double negative charge are
brought into a state of colloïdal suspension that appears in the fluid
in form of a black dye with a low settling rate of the suspended particles ($\sim$ 0,1 mm/s) and low rate of diffusion ($<$ 10^{-2} mm/s).

The tellurium wires or tellurium coated steel wire can serve for
100 experiments in laminar low speed flow.

2.6. Material tracer localized in a part of the flow

When vapor bubbles form and grow as a consequence of pressure
reduction, a liquid is said to cavitate. A two-phase flow, called cavita-
ting flow may be used as a flow visualization in the law-pressure regions
by reducing boundary pressure conditions. In circulating water channel
cavitation and supercavitation have been studied on marine propellers,
on hydrofoil, in the wake of profiles. Stroboscopic lighting shows the
individual bubbles of cavitation and the lowest pressure vertices.

If the experiment is got by cavitation similarity, the cavitation
number $\sigma = \dfrac{2 (p - p_c)}{\rho v^2}$ must be the same

when p   is the ambiant absolute pressure

po   is the cavity absolute pressure

v and ρ are velocity and density characteristic of the flow.

Mac Gregor (1961) produced screen of vapor in a supersonic wind
tunnel with running the tunnel with moist air : the air cools and the
moisture condenses to form a fog, like in the wake of wings of landing
planes because low pressure.

Another simple method to get information and visualization on some parts of a flow is to attach for instance some centimeters long tufted nylon yarns in a cross section or on the surface of solid body like on the sails of racing ships.

AIR FLOW INDICATION
ON SAILS BY MEANS
OF WOOLED TUFTS

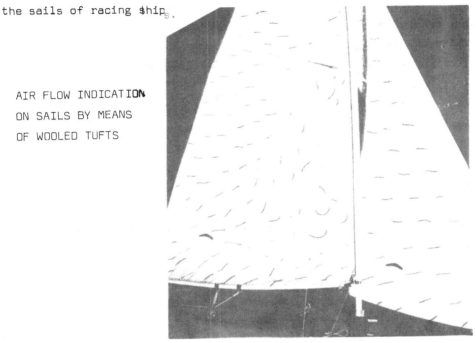

In aeronautical studies, with a tuft system on . rid places in a wind tunnel, quantitative measurements of the velocity have be performed by camera recording from downstream. Stiffness of the material and gravitation need an air speed above 20 m/s and a yarn length very smaller than the radius of curvature of the streamlines.

Close to the wall of a solid body, to avoid the pollution caused by a dye injection, and to allow a close-loop flow, one may coat discret dots or the whole wall, with a special material.

A picture of the flow pattern is given by the oil-flow technique in a steady flow : streaky deposits of the pigment powder can by photographed or studied on the body when the flow is stopped. The art of the experimenter is to find a paint of suitable consistency to leave a good photographic quality of the pattern, sometimes illuminated with ultra-violet light if fluorescent paint. Titanium dioxide, china clay, lamp-black and fluorescent chrysene are common pigments used in kerosene or similar oils. Discret dots allow quantitative measurements after cali-bration.

Like on sandy bottom of the sea, a relief pattern due to the pressure distribution occurs on a deformable or alterable surface body : a sooted glass plate shows shock waves (Mach 1878). This technique freezes a picture with criss-crossed pattern on the body. The atmosphe-ric reentry of space vehicles give cross hatching on ablating bodies, caused by the structure of shock waves above the surface.

Let us mention all the free surface visualization in fluid mecha-nics laboratory, by mean of spots, aluminium flakes, balls, highted-ships, confetti, etc... With help of camera and chronophotographic system thay may be very helpful both in stationnary and in unsteady flow.

3. ENERGY ADDITION USED AS A TRACER TECHNIQUE

If energy is transferred to certain fluid elements they can become tracers in the flow and be observed.

Incompressible flows get density variations by added energy at singular points. The altered density can be visualized, either by opti-cal methods (cf 4), or the luminosity of the fluid if a high rate of energy is used.

3.1. Adding heat

Electrical heated thin wires across the stream may give filament lines in a low speed air flow, because fluid particles have passed close to the wire and got raised temperature. The ratio of diffusive velocity to flow velocity must remain small enough not to disturb the main flow : in laminar flow, or to detect the transition condition, the method is appropriate with the help of an adequate optical technique.

A pulsed electrical current used for heating the grid of wires can get hot spots allowing a kind of mapping observation or photographs.

In water, thermal cloud given by a pulse can be followed by optical way due to change of refractive index (cf 3) or can be detected by electrical way due to conductivity change (Piquemal - Truchasson 1964).

3.2. Hot spot giving gas ionisation

Rarefied gases having a very high kinetic energy become luminous in the stagnation region for the kinetic energy is changed into heat enough to excite electronic transition in the gas so that a part of the flow is visible.

But in ordinary compressible flow, an intense spark discharge between two electrodes ionizes the gas. The lifetime of the luminous plasma is limited by the cooling process and is a function of the gas pressure. Direct observations of the luminous spark plasma or drop velocity recordings with high-speed camera, are mostly applied to the investigation of supersonic flows (Mac Intosh 1971). The spherical shock wave propagating from the spark reduces the accuracy of the velocity measurement, but can be observed by optical systems (cf 4). The observation of this blast wave motion in higher gas pressure and lower velocities in a flow without high velocity gradients (that would distort spherical cross sections) allow recording and velocity measurement of the "source center" (Miller 1967). To avoid the disturbance in the flow due to the electrodes a high energy pulsed laser beam can be used to get the ionised plasma drop (Chen 1966) (Leighton 1970).

3.3. Velocity napping by means of spark tracers

Because the first passage of a spark in a gas, gives an ionised column swept downstream, a second spark produced during the lifetime of the colums ($\sim$ 100 μ sec) choices rather this preionized path than the shortbest way between the electrodes.

An open-shutter photograph shows the successive locations of the ionised column illuminated in the flow by each spark.

One can get good mapping with two rod electrodes (Frungel 1970).

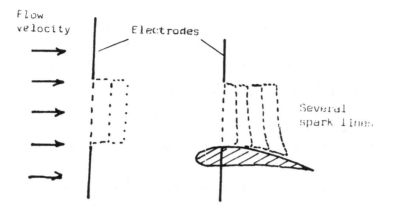

SPARK TRACER TECHNIQUE

3.4. Electron beams

In low density compressible flow, quantitative measurements and
flow visualization are got with a narrow beam of high-energy electrons
through the gas stream. Inelastic collisions between fast electrons and
the gas molecules, excite a great portion of these molecules that return
into the normal state with characteristic radiative emission of light
(Grün 1953). A spectroscopic quantitative analysis allows to detect
vibrational and rotational temperature and concentration rates of the
active gas (Muntz 1968).

A proportion of the gas particules are excited to a metastable
energy level the lifetime of which is relatively long, so that an afterglow radiation decreases when the particle is swept an increasing distance from the beam. The afterglow luminescent light is appropriate to
visualize shock waves in cold flows of nitrogen, argon or noble gases.
The afterglow is an error source in direct radiation recording.

Glow discharge have given good results in hypersonic nitrogen flow
and helium flow. The color and the intensity of the light from the positive column depends on the gas pressure (10^{-3} or 10^{-4} bar). Trepaud et
Brun (1967) have used high frequency field of about 20 MHz.

3.5. Chemiluminescence

A chemiluminescent reaction may occur when a two-body collision gives a chemical reaction with an emission of light, for instance, in air :

$$NO + O \longrightarrow NO_2 + h_{\nu}$$

This luminescent reaction exists when atomic oxygen and NO are present or formed by any mean, for instance by an electron beam or glow discharge in a low pression gas given by a 1000 V voltage between electrodes in air.

Hartmian and Spencer (1966) injected NO into a mean air flow through the walls of a porous model. Van der Blieck (1967) used the orifice of an injection nozzle. The separating streamline between injected NO and the free stream becomes very clearly visible. Such a visualization shows wakes, the points of separation and of transition from laminar to turbulent flow.

The mechanism occurs in rarefield gases and in some chemiluminescent solutions. Springer (1964) got a very high contrast of the glow in H_2O with KCl electrolyte and a chemiluminescent commercial product called LUMINOL ; platinium plated electrodes are about under 1 volt d.c. voltage It is possible to measure local or average heat and mass transfer rates on the anode model.

Bonneau (1977) excits ZnS phosphorescent particules by a stroboscopic laser beam ; he gets the velocity field in a thin film for lubrification by means of stroboscopic photographs.

PRINCIPLE OF BONNEAU' S
LUMINESCENT TECHNIQUE

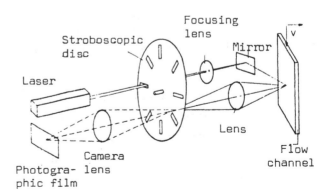

4. OPTICAL METHODS

When the flow under investigation is compressible, the fluid
density in the flow field can change either locally or with time. The
optical index of refraction (n = c/v) is a function of the gas density
so that a compressible flow field makes an optical disturbance for each
light ray passing through this field. On the contrary, we may assume that
the light disturbance on the flow is negligible.

According to the Biot and Arago's law((n-1)/ρ = constant) if the
density change (ρ (M,t)   ) is get isentropically (kinetic energy $\rightleftharpoons$
potential energy), the local change of density is a function of Mach
Number.

In fact, optical index method is available in a gas flow with
Ma $>$ 0,2 to produce a noticeable change in density, without external
heating or internal chemical reaction. But some degrees of temperature
difference, in a natural convective flow, produce noticeable density
variation.

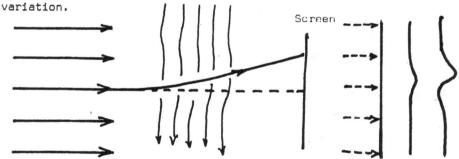

Deflection due to   Waves surfaces
optical disturbance   giving retardation

DEFLECTION AND RETARDATION OF LIGHT RAY BY AN OPTICAL DISTURBANCE

All optical methods use disturbance due to the inhomogeneous distribution
of the refractive index in the flowing gas : either the ray is deflected
from its original direction, or the phase of the disturbed ray is shifted
with respect to the undisturbed one. If we have exact knowledge of the
relation between optical change and density of a gas, a quantitative

density measurement can be performed.

4.1. The shadowgraph

Anybody has seen light moving shades in cold water basin into which a hot waterjet flows. Without any special device, that is an elementary sample of the shadow effect due to the thermal change of optical index n.

According to the Descartes'lax the angular deviation of a light beam is given by :

$$d\vec{\theta} = d\vec{s} \wedge \frac{\vec{\nabla} n}{n}$$

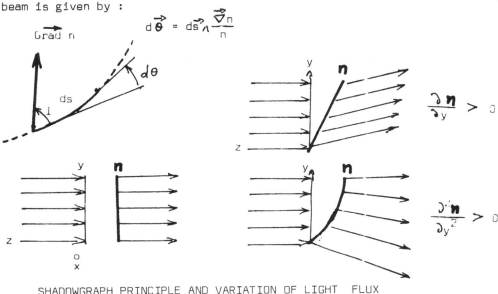

SHADOWGRAPH PRINCIPLE AND VARIATION OF LIGHT FLUX

Also, if a parallel light beam (z'z) passes through a compressible flow (x'x) between two parallel plane windows (xoy), the inhomogeneous distribution of the refractive index occurs deflection of the beam. The picture on a screen (XOY) of the wind tunnel flow shows a density distribution of illumination : for an index only function of y, in the place where dn/dy is a positive constant, any light beam is shifted equally upward (no luminous effect) ; if $\partial^2 n / \partial y^2$ is a positive constant the light flux on the picture is reduced. According to the $\partial^2 n / \partial y^2$ and $\partial^2 n / \partial y^2$ values, brighter and darker regions (more sensitive for human eye) appear. The shadow of a shock wave shows a dark line followed downstream by a sharp bright line due to an optical caustic.

Because of these assumptions and the summation along the beam, this method is roughly sensitive to the second derivative of the density and is unable to give quantitative measurement of the density in the space field.

Nevertheless, the shadowgraph is a simple cheap and convenient technique getting a quick survey of the flow pattern in steady or unsteady flow, specially for visualization of two dimensional shock-waves. Moreover, the image of two wires taped on the exit of the window is broken through a shock-wave : the distorsion of a wire is a measuring tool related to the change of density.

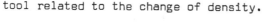

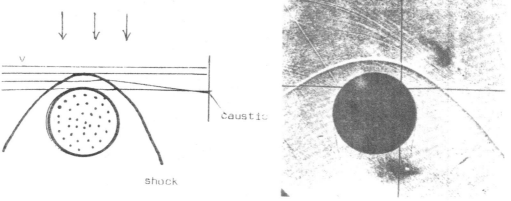

OPTICAL EFFECT OF A SHOCK-WAVE

The density fluctuation in turbulent compressible flow of a supersonic projectile has been studied by short duration light pulses recording : the fluctuating density acts as very small lenses that change in shape and position, the average of which can be correlate with the photographic pattern (Uberoi and Kovasznay 1955 ; Taylor 1970).

4.2. Princip of the Schlieren method

A simple optical arrangement is able to give good optical visualization with a high degree of resolution (Foucault 1859 ; Toepler 1964) : it is known as the Schlieren method.

Let us consider a light source LS illuminating the test section with parallel beam through an optical system that gives the image L'S'. If a solid precise mask cuts off the light source image, dark is the screen on which appears the image of the flow section without disturbance of optical index. But any density disturbance in the test section deflects the light ray and illuminates a part of the screen.

PICTURES OF THE SAME STEADY GAS FLOW
GOT BY SHADOWGRAPH   AND   BY SCHLIEREN TECHNIQUE

If a knife edge (ox) cuts off a part of the light source (either point or narrow slit parallel to the knife edge), the illumination changes $\angle \dot{I}$ give dark and bright zones due to $\partial n/\partial y$ accordint to the shift

With turning the knife edge by 90° one can get the same information about $\partial n/\partial x$. But diffraction restricts in reducing the aperture   and this simple method is mostly employed for qualitative purposes, and other systems and modifications are more and more frequently used.

4.3. Varied schlieren systems

A double pass schlieren systems uses a mirror behind the flow so that the sensitivity is better.

Rectangular double knife edge, circular light source and smaller circular mask are also available. A moiré pattern of fringes suitable

for quantitative evaluation is observed with a grid light source LS and
a parallel grid in L'S'

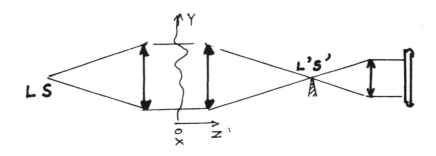

ELEMENTARY SCHLIEREN SYSTEM

Schardin (1942) used a crossed slit method and Knoos (1968) the
astigmatic aberration of an off-axis device with a spherical mirror.
Three dimensional resolution has been described including a plane distribution of multiple point sources and knife edges (Kantrowitz and Trimpi
1950).

But a convenient and beautiful devise is easy to get with a color
schlieren systems.

Instead of a knife edge, in the focal plane of the converging lense
one puts a coloured filter G made of several strips parallel to the
light source slit.

In a constant density flow all the screen is coloured by the central strip. Any disturbance deviates the ray which gives another colour
through the striped grid and enables quantitative measurement on

$$\int_{z'z} \frac{\partial n}{\partial y}\ dz$$

Because the eye seems more sensitive to change in colour than in
shades of gray, this technique give good flow visualization of the index
variation. Like in the Toepler method, turning the coloured grid by 90°
gives information on $\partial n, \partial x$.

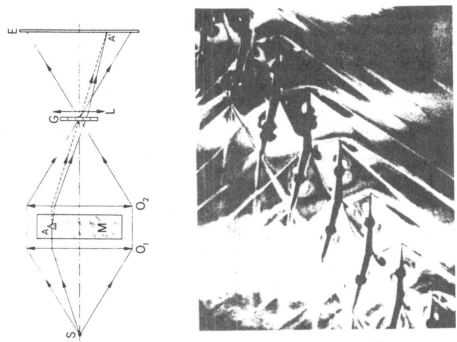

COLOUR   STRIPED **TECHNIQUE IN SCHLIEREN METHOD**
(From   ONERA)

This visualization is very closed to the hydraulic analogy with a
free surface water table (5.3) with a colour striped bottom and deviation
of the beam.

The same kind of show is obtained with an interferometric device
using a Wollaston prism.

4.4. Interferometric methods

Many a device is available. A phase alteration of the light through
the flow field is exploited for producing or changing a fringe pattern.

The coloured pattern obtained with a Wollaston prism in the position L'S' of the optical arrangement of the Schlieren system, are made of
coloured fringes due to the flow disturbances that deviate the light beam.

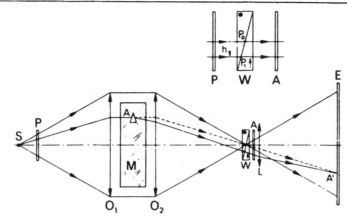

INTERFEROMETRIC METHOD USING A WOLLASTON PRISM

A Wollaston prism is made of two prisms P_o (crystal axis ox) and P_1
(crystal axis oy). The incident light is polarized along 45° direction by
passing through a polarizer P ; the emerging light is analysed through the
crossed polarizer A. Because of the Wollaston double prism, interferometric
fringes appear on the screen E, without disturbance in the flow field M.
Any disturbance turns the ray so that the colour changes on the corresponding spot. The deviation is sensible to $\int \frac{dn}{dy}$ dz if ox is the direction
of the sharp edge of the prisms which act like a birefringent model of
various thickness according to the beam deflection. Monochromatic light
changes the coloured strip into dark and bright fringes.

            The phase shift must be measured for quantitative experi-
ments. It is better to shoot a first photograph of the fringes pattern
without flow then the exprimental pattern, to eliminate optical disturbances. White illumination is used for localization of the order zero fringe.
Monochromatic illumination and low variation of the speed in the wind-
tunnel allow progressive and quantitative measurement of the shift, then
calculation of the density related to the velocity calculation.

            This excellent and precise technique has been developped
by the introduction of laser holograms.

    4.5 Holographic interferometry in flow visualization :

    In the interferometric method used in flow visualization, a laser
beam can be used for monochromatic illumination.

Another possibility is to redord an hologram H on the same photographic plate with the first exposure (E) without flow and the second one
(E') with the gas flow.

The laser restitution shows the only fringes generated by the gas
flow. Ruby light pulsed laser (6943 A) in the manosecond range, can serve
for high speed photography (cf 6.2.), but high energy laser, like
Q-switched laser ($\sim 10^9$ W) are not always undisturbing investigators
(cf. 3)

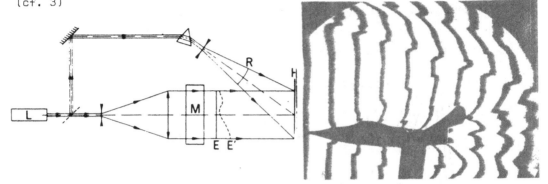

HOLOGRAPHIC INTERFEROMETRY

5. ANALOGICAL SYSTEMS IN FLOW VISUALIZATION

Sometimes the flow visualization is transposed on another fluid and
on a scale model, because much more convenient to experimentation.

5.1. Streaming birefringence

Interferometric pattern close to the results ~iven by the precedent
techniques (4.4) of visualization are available with some incompressible
fluid flows. In birefringent fluid the optical index is not a function
of the density but a function of the velocity field through a rheo-opti-
cal law.

In a birefringent fluid at rest, the molecules are distributed at
random. They may be aligned by the action of an electrical (Kerr effect)
magnetical or mechanical external stress. This optical anisotropy may
appear too if the fluid contains molecules which are deformed or oriented by shear stresses, for instance.

Between the two crossed polarizers, an interference pattern, (isochromates and isoclines) can be recorded, due to the interference of the phase shifted ordinary and extraordinary light waves.

Elimination of isoclinic fringes is obtained by classical optical fourth wave length sheets, like in photoelastic technique. A white light source shows coloured visualization, but precise measurement needs monochromatic light the reponse of which must be studied according to the fluid and the detection system (Pindera 1973)

Some liquids and solutions become more or less double refracting under shearing motion, but the behaviour may be Newtonian or non-Newtonian according to their nature and their physical parameters.

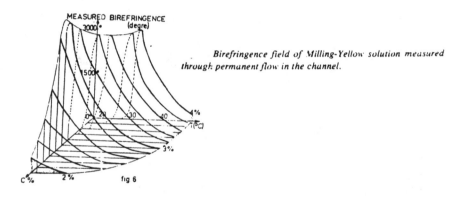

Birefringence field of Milling-Yellow solution measured through permanent flow in the channel.

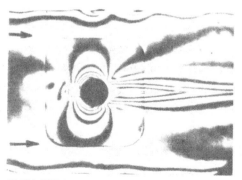

FLOW BIREFRINGENCE IN A STEADY FLOW
AROUND A CIRCULAR CYLINDER (from RANDRIA)

A very high birefringence is given by a solution of milling yellow dye in distilled water. A reversible behaviour after thermal and mechanical stresses, no ageing effect and very clear solution can be obtained by an adequate preparation of the solution (Jibawi 1975). The high sensitivity of such a solution to the temperature and the concentration are due to the structure of the fluid. Quantitative measurements are available with Newtonian behaviour corresponding to low concentration (below 1%), but low double refraction effect.

The best use is for visualization of flow with laminar and turbulent parts (the birefringence vanishes when the turbulence rises) wakes problems and unsteady flows in scale models (Randria 1977).

Three-dimensional study could be performed by adapted optical technique (Pindera 1974).

5.2. Similitude analogy

In a laboratory, it is often more convenient to study a scale model with another fluid, because of the boundary conditions, of the velocity or the pressure field or the time scale due to the similitude laws.

The flow visualization, in respect with the Reynolds number is convenient in an hydrodynamic tunnel for gas flow at low Mach number. In combination with dye injection this technique is often used in aerodynamic research (Werle 1974) and atmospheric studies.

Sometimes, the flow visualization takes place in a wind tunnel for studying problems of fluid flow in charge (submarine flows).

5.3. Hydraulic analogy with a free surface water table

A two dimensional steady isentropic compressible gas flow is analogous to a steady unviscid incompressible free surface water flow.

A rigid body profile in a water channel shows thereby wave patterns similar to the supersonic waves distribution.

The liquid depth h must be smaller ($\sim 0,5$ cm) than the wave length of the surface waves ; the slope of the water surface has to be low. The boundary effect caused by the viscosity and capillarity give disturbances on the thickness of the shock wave and shows secondary waves.

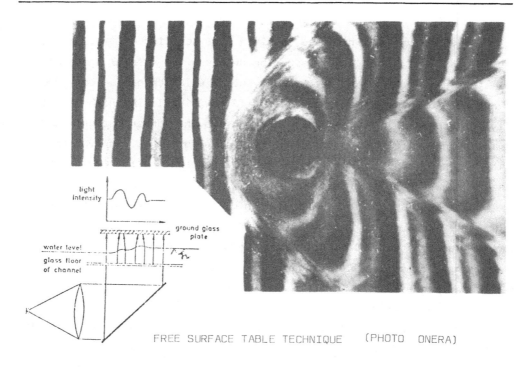

light
intensity

ground glass
plate

water level

glass floor
of channel

FREE SURFACE TABLE TECHNIQUE   (PHOTO ONERA)

Nevertheless this hydraulic analogic has been used with success
to investigate for instance supersonic flow through turbomachines, sometimes in combination with a tracer particle technique (Le Bot et Bernard
1972).

Unsteady wave patterns give a convenient and valuable visualization by means of the hydraulic analogy. Because of the speed ratio of the
sound in air, to the gravity waves in water, the entire process is slo-
wed down from 1000 to 1 in the water channel. The flow visualization does
not need any high-speed photographic equipment but can be helped by optical gadgets.

The show visualization is more beautiful and seems to be more
precise if the bottom of the channel is made of parallel coloured rays.
Because the slope of the free surface deviates the light beam, any
change of depth h gives a possible colour change (cf. 4.3)

To obtain the most pleasant effect, the wideness and the direction of the strips depend upon the surface wave situation.

5.4. Hele-Shaw analogy

By means of coloured liquid injection, the visualization of a creeping motion in steady flow gives the filament lines of a plane instational incompressible flow : that is the Hele Shaw-viscous flow analogy.

A creeping motion is a fluid flow in which inertial forces can be neglected in regard to viscous forces. The Reynolds $Re = \frac{VL}{v}$ number must be much smaller than 1.

The stream functions (ψ = C along a streamline) of a creeping motion must verify the same differential equation as do the stream functions of a potential inviscid and incompressible fluid.

As the limit of the flow is a particular streamline, in steady flow the filament lines are the same. But their "graduation" is different the velocity distribution is not the same because the boundary conditions are different ($\vec{v} = \vec{0}$ in creeping motion ; $\vec{v}.\vec{n} = 0$ in inviscid flow).

Of course, the both flow must be laminar. This condition and the creeping motion requires a very narrow spacing between the plates, of the order of 1 mm when heavy oils or glycerine are used.

The density ρ and the viscosity v must be constant and the temperature controlled : in some occasions it is necessary to cool the pumped fluid after it leaves the pump.

Many a teaching experiment or a research visualization for infiltration problem has been showed by the Hele Shaw method.

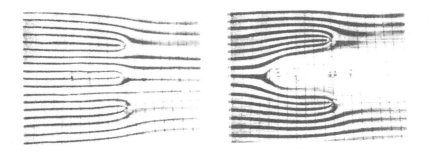

HELE-SHAW ANALOGY FOR A STEADY FLOW

6. RECORDING AND ILLUMINATION TECHNIQUE IN FLOW VISUALIZATION

The choice of illumination and recording system depends on the fluid velocities and on the kind of investigation.

A plane sheet of light may be sufficient for some qualitative studies. Sometimes, a quantitative measurement is got by stroboscopic lighting.

A quick phenomena requires high speed photography that can be combined with microscopic techniques.

Out of the visible range, illumination or recording is able to get convenient information specially about mixture problems with electronic camera and T.V. recording set.

6.1. Illumination technique

With local injection of tracers, like boundary coloured liquid injection (2.1) or local energetic injection (3), a whole illumination of the field is necessary. That one is also used in many analogous visualization methods (5).

Hologram of the whole illuminated flow gives a true three-dimensional image that allows the reconstruction of the tracers in a thin plane sheet with a short focusing observation system.

When tracer particles are suspended all over the field a plane examination is allowed by a thin plane slice illumination of the flow. Diaphragm and convenient lenses can give good results with careful experimentation : narrow ribbon of light from a parallel laser beam, coloured parallel sheets for three dimensional determination (Van Meel and Vermij 1961), etc...

The particles out of the illuminated slice disturb the recorded information. Nevertheless the thin sheet illumination is a very helpful technique for qualitative study of vortex and velocity distribution.

Quantitative measurments need a well known shutter-time of the camera. The velocity vector must be in the lighted plane sheet; any normal component is a source of error.

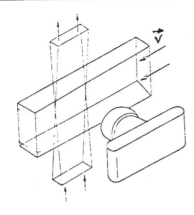

PLANE SHEET OBSVERVATIUN (PHUTU UNERA)

Because camera shutters do not operate with high precision, the particle streak can be determined by the duration of a light flash ; the most accurate process seems to be the generation of short lights pulses with an electronic delay time between pulses. Philbert and Boutier (1972) used two time delayed ruby lasers with light pulses of 2×10^{-8} sec. Though expensive, it shows a spectral homogeneity with a very concentrated luminous center.

It is better than a single electric spark, because the luminosity is not constant along the flash duration : the light emission reaches its maximum very quickly but drops slowly to an afterglow about 10% of the maximum value, even in Ar, Kr, Xe giving the greatest light amplitude.

An electro-optical shutter shorter than 1   sec can be got by a Kercall. Nitrobenzol for instance, becomes birefringent when exposed to an electric field so that a part of the light is transmitted by a vessel fluid between crossed polarizers. Without field quite no light is transmitted . Some kilovolt voltage are required and the optical path causes an important loss of light intensity.

Electro-optical crystals and liquid crystals could be used too.

Formerly, stroboscopic illumination has been often experimented in combination with plane sheet slice or special camera technique. The camera can be in open flash position with a permanent illumination and a rotative disk with holes either in front of one camera lens, or between the light system and the fluid flow. This technique have done good results, and is not too expensive. It has been very convenient for examination of cavitation structures in turbomachines.

We do not tell about the electronical detection technique proposed by Gaster (1964) because like with the doppler velocimeter (Kumarakulasinghe, 1973) one can measure a velocity without any flow visualization.

6.2. Special photographic techniques

Out of the techniques of illumination, special camera sevices can be used.

A microscope gives a large picture of a small flow field. The light must be intensive and often it is obtained by electronic pulses synchronised with the camera shutter ; it is necessary when a high-speed cinematography is used for resolving events that change quicker than the reaction time of the eye.

For instance detonation waves travelling with a 500 m.s^{-1} speed, provide on exposure time of 10^{-6} seconde for good focusing. If the observer of the movie wants to see a 10 cm.s^{-1} speed, with a 20 pictures per second film, the movie camera must record 10^5 pictures per second : a 5 second experiment requires about a 4 km long film (in 16 mm size). In spite of the price oe this process, such a visualization is used up to exposure time below 10^{-6} second, with associate illumination system (cf. proceedings of the International Congress on high-speed photography).

The total velocity of a film restricted to the greatest achievable frame rate of 100 pictures/sec. if the film is stopped for the duration of each single exposure. With a continuous motion of the film, the separation of each frame can be achieved either by a stroboscopic illumination of the flow, or by a system of rotating mirrors or rotating prisms : each single picture moves along with the film at the same speed. The time interval of coincidence is the exposure time of a single picture. A rotating prism allows about 10 000 frames/sec. In combination with drum camera the greatest possible frame rate is about 50 000 frames/sec.

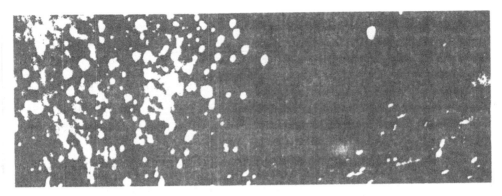

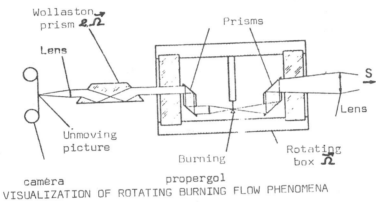

Wollaston prism ℓ, Ω
Prisms
Lens
S
Lens
Unmoving picture
Burning
Rotating box Ω
caméra
propergol
VISUALIZATION OF ROTATING BURNING FLOW PHENOMENA

Sometimes the camera system is complicated by technical agencies, like in the visualization of the burning flow in a rotative rocket.

We have seen how polarized light can help in flow visualization to make visible what is not by means of birefringent flow (5.1.)

Electronic camera can be used too, to make visible what cannot be seen by human eyes, for instance through X-rays of infrared radiation. Electronic colour camera is convenient for studying mixture problem in free surface flow and for teledetection or petroleum pollution in lakes and seas.

C O N C L U S I O N

The flow visualization may be a direct method such as an optical undisturbing system (4), an analogical method such as the hydraulic analogy with a water table (5.3) or an indirect method in which one observes the motion of foreign material (2) instead of the fluid itself.

For quantitative measurement, the difference between the movement of the fluid and that of the foreign particles must be minimized.

But the choice of a technique depends not only upon the required precision and upon the problem, but alson upon the hardware and software investment.

Some methods need an advanced technology and in fact are reserved to the top research laboratory (energy addition 3) or need a very specialised team (holographic interferometry 4.5)

Other techniques are easy to use and some of them are cheap : coloured tracers (2.1), smoke visualization (2.2). Stroboscopic devices are not very expensive. The hydrogen bubble technique (2.4) can be achieved with an electrical simple apparatus.

When a transonic or supersonic wind tunnel is not disposable, the free surface table water (5.3) is able to give a cheap qualitative information. Out of large technical surrounding, a birefringent solution allows unsteady studies (5.1) in close loop.

All the techniques we mentionned can be more or less interesting according.to the nature of the problems and according to the requested informations.

A large variety of flow visualization experiments have been performed, and no doubt that new techniques will arise in combination with associated quantitative measurement.

Owing to the technical progress reducing the physical size of the measuring probes, and the temptation of systematic computing treatment of some punctual measurements, the techniques of flow visualization could be misregarded.

Let us forget the intensive interest of a good visualization, with the pleasure of the show, its whole or large field of observation of the flow, its instantaneous reality, and all its subjective and implicit informations. Many an old technique can be used twice : the first as an instantaneous observation, the second for recording with all the present systems for quantitative squeeze even with computer calculation.

Specially in research report and in teaching the flow visualization is an irreplaceable technique, covering a wide field of scientist interest

Various books and references contain precise and precious technical details with theoritical approach and particular possibilities of application in special situations.

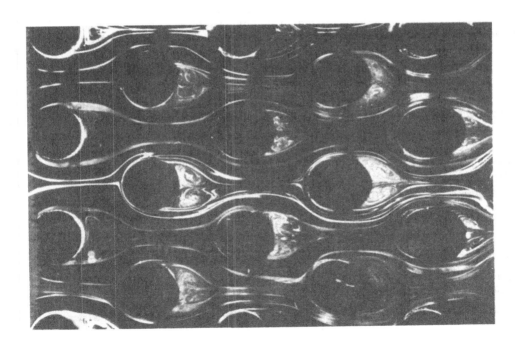

R E F E R E N C E S

MAIN REFERENCE : W. MERZKIRCH . Flow visualization (1974)

- A. BOUTIER, M. PHILBERT, J. SURGET, C. VERET ; Visualisation et procé-
dés optiques de mesure en aérodynamique - Techniques de l'Ingénieur -
R 2160 (1978)

- P. BRADSHAW . Experimental fluid mechanics (1970)

- A. FORTIER . Mécanique des suspensions (1967)

- P. REBUFFET . Aérodynamique expérimentale (1950)

- O.N.E.R.A. Publications internes (Chatillon - France)

- BECKER H.A., HOTTEL H.C. and WILLIAMS G.C. . On the light-scatter
technique for the study of turbulence and mixing - J. Fluid. Mech. 30,
259- 284 (1967)

- D. BONNEAU and J. FRENE . Phosphorescent tracer method to measure
velocity profiles in fluid film lubrification - Tribology international
June 1978 - p. 181-184.

- BROWN F.N.M. . A photographic technique for the mensuration and evaluation of aerodynamic patterns . Photog. Eng. 4, 146-156 (1953)

- C. CAMICHEL . Sur la détermination des vitesses dans les liquides -
Revue générale de l'électricité - Paris - 22 novembre 1919 - Tome VI -
p. 107-109.

- C. CAMICHEL, L. ESCANDE . Similitude hydrodynamique et technique des
modèles réduits - Publication Scientifique et Technique du Ministère de
l'Air - n° 127 - Paris (1938).

- CERESUELA R., BETREMIEUX A. and CADARS J. (1965) . Mesure de l'échauf-
fement cinétique dans les souffleries hypersoniques au moyen de peintures
thermosensibles. Rech. Aérosp. n° 109, 13-19.

- CHEN C.J. (1966) . Velocity profile measurement in plasma flows using
tracers produced by a laser beam - J. Appl. Phys. 37, 3092-3095.

- CLAYTON B.R. and MASSEY B.S. (1967) . Flow visualization in water :
A review of techniques - J. Sci. Instrum. 44, 2-11

- CLUTTER D.W. and SMITH A.M.O. (1961) . Flow visualization by electrolysis of water - Aerosp. Eng. 20, 24-27, 74-76.

- CORCORAN J.W. (1967) . Applications of the isodensity tracer in high
speed photography in "Kurzeiphotographic (Proc. 7th Int. Congress on

High-Speed Photography)" (O.Helwich, ed.), pp. 466-471. O Helwich, Darmstadt, Germany.

- COUTANCEAU M. . Sur la détermination des principales caractéristiques des champs hydrodynamiques au moyen de la visualisation par traceurs solides - Colloque de l'A.R.T. - Paris 1978.

- CROWE C.T. (1967) . Drag coefficient of particles in a rocket nozzle - AIAAJ. 5, 1021-1022.

- DANCKWERTS P.V. and WILSON R.A.M. (1963) . Flow visualization by means of a time reaction - J. Fluid Mech. 16, 412-416.

- DENBIGH K.G., DOMBROWSKI N., KISIEL A.J. and PLACE E.R. (1962) . The use of the "time reaction" in residence time studies - Chem. Eng. Sci. 17, 573.

- FOUCAULT L. (1859) . Ann. Observatoire Paris 5, 197.

- FRANCON M. (1952) . Interférométrie par double réfraction en lumière blanche - Rev. Opt., 65-80, n° 31.

- FRÜNGEL F. (1970) . Messung der dreidimensionalen Lufwirbelung nach dem sog spark-tracing Verfahren - VDI (Ver. Deut. Ing.) Berichte n° 146.

- GASTER M. (1964) . A new technique for the measurement of low fluid velocities - J. Fluid Mech. 20, 183-192.

- GODDARD V.P., McLAUGHLIN J.A. and BROWN F.N.M. (1959) . Visual supersonic flow patterns by means of smoke lines - J. Aerosp. Sci. 26, 761-762

- GRUN A.E., SCHOPPER E. and SCHUMACHER B. (1953) . Electron shadowgraphs and afterglow pictures of gas jets at low densities - J. Appl. Phys. 24, 1527-1528.

- HARTUNIAN R.A. and SPENCER D.J. (1966) . Visualization technique for massive blowing studies - AIAAJ. 4, 1305-1307.

- HJELMFELT A.P. and MOCKROS L.F. (1966) - Motion of descrete particles in a turbulent fluid - Appl. Sci. Res. 16, 149-161.

- JIBAWI M.H. . Influence de certains paramètres physiques sur la biréfringence d'écoulement - Thèse n° 475 - Toulouse 1975.

- JOUGUET E. (1920) . Quelques problèmes d'hydrodynamique générale - J. Math. Pures Appl. 3, 1-63.

- KANTROWITH A. and TRIMPI R.L. (1950) . A sharp focusing schlieren system - J. Aero. Sci. 17, 311-314.

- KNÖÖS S. (1968) . A quantitative schlieren technique for measuring one-dimensional density gradients in transparent media - In "Proc. 8th Int. Congress on high speed photography" (N.R. Nilsson and L. Wögberg eds), pp. 346-350 Almquist & Wiksell, Stockholm.

- KUMARAKULASINGHE R.B. . Utilisations d'un anémomètre à laser en mécanique des fluides - Thèse n° 1474 - Toulouse 1973.

- J.P. LALLEMAND, R. DESAILLY, C. FROEHLY . Mesure des vitesses dans un
liquide par diffusion cohérente - Acta Astronautica vol. 4, 343-356,1977.

- LEIGHTON R.L. (1970) . Gas velocity measurement employing high-speed
schlieren observation of laser-induced breakdown phenomena - In "Proc.
9th Congress on high-Speed Photography" (W.G. Hyzer and W.G. Chace, edt)
pp. 93-96. SMPTE, New York.

- LE BOT Y. and BERNARD P. (1972) . Analyse par analogie hydraulique du
décollement tournant dans les compresseurs. - Rech. Aérosp. n° 1972-4,
187-198.

- MACH E. and Von XELTRUBSKY J. (1878) . Über die Formen der Funkenwellen
Sitzungsber. Kaiserl. Akad. Wiss. Wien, Math - Naturwiss. Kl. Abt. 1, 78,
551-560.

- MACH L. (1896) . Über die Sichtbarmachung von Luftstromlilien - Z. Luft-
schiffahrt Phys. Atm. n° 6, 129-139.

- MALTBY R.L. and KEATING R.F.A. (1962) . Smoke techniques for use in
low speed wind tunnels - AGARDograph n° 70, 87-109.

- McINTOSH M.K. (1971) . Free stream velocity measurements in a high
enthalpy shock tunnel - Phys. Fluids 14, 1100-1102.

- MILLER H.R. (1967) . Shock-on-shock simulation and hypervelocity flow
measurements with spark discharge blast waves - AIAAJ. 5, 1675-1677.

- MUNTZ E.P. (1968) . The electron beam fluorescence technique -
AGARDograph n° 132.

- Mc GREGOR I. (1961) . The vapor-screen method of flow visualization -
J. Fluid Mech. 11, 481-511.

- PHILBERT M. et BOUTIER A. (1972) . Méthodes optiques de mesure de
vitesses de particules entraînées dans les écoulements - Rech. Aérosp.
n° 1972-3, 171-184.

- J.T. PINDERA and A.R. KRISHNAMURTHY . Foundations of flow birefringence
in some liquids - Experimental Mechanics in research and development
p. 545-599 - Waterloo, Canada - 1972.

- P. RANDRIA . Biréfringence d'écoulement en mécanique des fluides -
Thèse n° 18 - Toulouse 1977.

- RIABOUCHINSKY D.P. (1932) . Sur l'analogie hydraulique des mouvements
d'un fluide compressible - C.R.H. Acad. Sci. 195, 998-999.

- SCHARDIN H. (1942) . Die Schlierenverfahren unde ihre Anwendungen -
Ergeb. Ewakten Naturwiss. 20, 303-439.

- SIGLI D. . Sur une méthode de détermination des caractéristiques de
certains fluides non newtoniens basée sur des mouvements oscillatoires

Application à l'écoulement autour d'un obstacle sphérique en milieu
limité - Thèse de Doctorat d'Etat - Poitiers 1976.

- SPRINGER G.S. (1964) . Use of electrochemiluminescence in the measurement of mass transfer rates - Rev. Sci. Instrum. 35, 1277-1280.

- Vom STEIN H.D. and PFEIFFER H.J. . Investigation of the velocity
relaxation of micron-sized particles in shock waves using laser radiation
Appl. Opt. 11, 305-307.

- TAYLOR L.S. (1970) . Analysis of turbulence by shadowgraph -
AIAAJ. 8, 1284-1287.

- TOEPLER A. (1864) . Beobachtungen nach einer neuen optischen Methode.
Max Cohen u. Sohn, Bonn.

- TREPAUD P., PERY R., GOEHLER J.P., VIVIAND H. and BRUN E.A. (1967).
Etude de sillage de cylindres et de dièdres en écoulement de gaz raréfié
AGARD Conf. Proc. n° 19.

- UBEROI M.S. and KOVASZNAY L.S.G. (1955) . Analysis of turbulent
density fluctuations by the shadow method - J. Appl. Phys. 26, 19-24.

- VAN DER BLIEK J.A., CASSANOVA R.A., GOLOMB D., DEL GRECO F.P., HILL
J.A.F. and GOOD R.E. (1967) . The chemiluminescent reaction of NO with O
in a supersonic low density wind tunnel - In "Rarefied Gas Dynamics",
5th Symposium (C.L. Brundin, ed), pp. 1543-1560. Academic Press, New York

- VAN MEEL D.A. and VERMIJ H. (1961) . A method for flow visualization
an measurement of velocity vectors in three-dimensional flow patterns in
water models by using colour photography - Appl. Sci. Res. A10, 109-117.

- WERLE H. (1960) . Etude effectuée à la cuve à huile et au tunnel
hydrodynamique à visualisation de l'O.N.E.R.A. - Rech. Aér. n°79, 9-26.

- WERLE H. and GALLON M. (1969) . Cuve hydraulique pour la visualisation
des écoulements au point fixe - Rech. Aérosp. n° 129 - 13-18.

- WERLE H. (1974) . Tunnel hydrodynamique au service de la recherche
aérospatiale - Publication 156 ONERA.

- WORTMANN F.X. (1953) . Eine Methode zur Beobachtung und Messung von
Wasserströmungen mit Tellur - Z. Angew. Phys. 5, 201-206.

Printed in the United States
By Bookmasters